Suppose

Robb Skoef

TABLE OF CONTENTS:

PREFACE

Some book authors, I'm taken to understand,
Write Prefaces and Introductions with pen in hand.

But I'm not an author, that's clear to me,
I'm a High School graduate and a railroad retiree.

That sums up my life,
I've not even a wife.

But then one day back in eighty-eight,
While contemplating the air and the sidewalk slate.

Some weird, thought ideas, in my mind seemed to bubble.
My God, was my thought, I've a sick mind in trouble.

I must say that trouble these thoughts seemed to be.
I just could not understand why they'd come to me.

I purchased some books 'bout forty in number,
And pondered weird thoughts from awakening to slumber.

I'd found a great truth; it seemed to me,
That could unlock life's' mystery.

But how hap'd I 'pon this magnetudenal truth,
Since the educated men had not it deduced.

I read thousands of words 'bout truth men had written,
And knew in my heart they couldn't chew what they'd bit-
ten.

They all claimed to be men of legitimate science,
But with one theorist they compounded an unholy alliance.

They all, by the hundreds used their facts, it now seems,
And bent them and twisted them to fit Darwin's Dreams.

Now Darwin had right to say just what he thought,
His words were then to his students taught.

But science, since eighteen hundred and fifty nine,
Has changed its factual endeavors to a theorist's line.

They've written it down in their Encyclopedias,
And expounded it on paper, radio and the TV's intellectual medias.

They gave us the facts, then comes what is eerie,
They want us to know that their facts fit the 'Theory".

Now I've only gone to school for a total twelve years.
But through all of my life I've used my eyes and my ears.

The one thing I know that science seems to lack,
Is the necessity to fit the theory, to the fact.
Thousands of theorists and a hundred thirty years of time,
Can't demonstrate that their theory fits better than mine.

By using the many facts these scientific intellects wrote,
I'll show you a theory that fits, like a silk scarf 'round a throat.

Like David facing Goliath I'll aim just 'bove the nose,
I feel like the child who said, "The King has no clothes."

Will I be laughed at and ridiculed for the bold things I speak,
Will I be punished and scolded for not being meek.

Will this Theory harm mankind or will it lead to advance.
I feel so it's truth, that I must take the chance.

So now if you'll join me two things you will need,
The desire to SUPPOSE and an open mind to feed.

I've wondrous thought and grand conclusions,
Science will say they're my mind delusions.

I don't mind that to themselves they were unjust
But I hate that they tricked me, they tricked us.

All life was pre-planned from protoplasm to sod,
All life was pre-planned by whom we call God.

That was a little ditty I wrote back in 1989, the year I retired. At that time I was angry at science and was setting out to write a scientific paper. Even more than that I thought I was going to write a scientific book, like Darwin's

HOW IT STARTED

I was sitting out in the back yard reading Rachel Carlson's book The Sea Around Us when, what I later began to call The Suppose Theory, first entered my mind. I'm not any kind of a scientist and truly have no right to even consider developing a Scientific Theory. Of course I never in a million years would have even thought of such a thought. There I was just enjoying this spring day reading and daydreaming about next year's spring and then my 60th birthday in July and my retirement after 42 years on the Railroad.

I was enjoying Rachel's book, as it was the type of reading that I have always enjoyed. It was non-fiction, informative and thought provoking. For most of my life and just for my own satisfaction I've been sort of a searcher for the truth. I did a lot of pondering but did not uncover any astounding truths. I Suppose in reality I was more of a wonderer than anything else. I'd read that the first seas were red and I wondered why. Science told me that India broke off from the Antarctic and moved "very rapidly" north slamming into Asia, I wondered why. In the sixth grade they told me about the "fossil" fuels, coal and oil and I wondered a lot about that. I wasn't able to comprehend how coal and oil could be so centrally situated when the dinosaurs and other dead animals and flora didn't all go to that spot to die. I could not fathom the Evolution Idea. I wondered how such a variety of complex structures had created themselves with such perfection, just by chance. I wondered why we could say; He looks Irish, or Polish, or German, or Chinese, Japanese or Korean. I wondered how one group of people had developed a guttural language and their neighbors a nasal language.

Anyhow I did a lot of wondering and was enjoying this section of Rachel's book because she was also wondering about her science and looking for truths. She spoke of the submarine canyons as one of the mysteries found below the waves. She spoke of sea mounds, just big piles of dirt. And she spoke of under the Sea Mountains which had their tops cut off. Like thousands of other people who had read these words I turned the page and continued reading. My slow High School educated brain was still pondering the last page and the way brains do it called me back to reread. Yea, yea, big canyons under the sea and large piles of Earth and mountains with their tops cut off, yea, yea. Lets see she was speaking about the slopes of the continental shelf and then she said:

> "The slopes are the site of one of the most mysterious features of the sea. These are the submarine canyons with their steep cliffs and winding valleys cutting back into the walls of the continents. The canyons have now been found in so many parts of the world that when sounding have been taken in presently unexplored areas we shall probably find that they are of world wide occurrence. Geologists say that some of the canyons were formed well within the most recent division of geologic time, the Cenozoic, most of them probably within the Pleistocene, a million years ago, or less. But how and by what they were carved no one can say. Their origin is one of the most hotly disputed problems of the ocean.
>
> Only the fact that the canyons are deeply hidden in the darkness of the sea (Many extending a mile or more below present sea level) prevents them from being classed with the worlds' most spectacular scenery. The comparison with the Grand Canyons of the Colorado is irresistible. Like river cut land canyons, sea canyons are deep and winding valleys, V-shaped in cross section, their walls sloping down at a steep angle to a narrow floor. The location of many of the larger ones suggests a past connection with some of the great rivers of the earth of our time. Hudson Canyon, one of the largest on the Atlantic coast, is separated by only a shallow sill from a long valley that wanders for more than a hundred miles across the continental shelf, originating at the entrance of New York Harbor and the estuary of the Hudson River. There are large canyons off the Congo, the Indus, the Ganges, the Columbia, the Sao Francisco, and the Mississippi, according to Francis Shepard, one of the principal students of the canyon problem. Monterey Canyon in California, Professor Shepard points out, is located off an old mouth of the Salinas River; the Cap Breton

> Canyon in France appears to have no relation to an existing river but actually lies off an old fifteenth century mouth of the Adour River.
>
> Their shape and apparent relation to existing rivers have led Shepard to suggest that the submarine canyons were cut by rivers at some time when their gorges were above sea level. The relative youth of the canyons seems to relate them to some happenings in the world of the Ice Age. It is generally agreed that sea level was lowered during the existence of the great glaciers, for water was withdrawn from the sea and frozen in the ice sheet. But most geologists say that the sea was lowered only a few hundred feet-not the mile that would be necessary to account for the canyons. According to one theory, there were heavy submarine mud flows during the times when the glaciers were advancing and sea level fell the lowest; mud stirred up by waves poured down the continental slopes and scoured out the canyons. Since none of the present evidence is conclusive, however, we simply do not know how the canyons came into being, and their mystery remains".

I had known about the submarine canyons but I had never heard of the sea mounts before. Here is what she said about them in her book published in 1951.

> "A new feature on recent maps of undersea relief-something never included before the 1940's-is a group of about 160 curious, flat-topped sea mounts between Hawaii and the Mariana's. It happened that a Princeton University geologist, H. H. Hess, was in command of the U. S. S. Cape Johnson during two years of the wartime cruising of this vessel in the Pacific. Hess was immediately struck by the number of these undersea mountains that appeared on the fathograms of the vessel. Time after time, as the moving pen of the fathometer traced the depth contours it would abruptly begin to rise in an outline of a steep-sided sea mount, standing solitarily on the bed of the sea. Un-

> like a typical volcanic cone, all of the mounts have broad, flat tops, as thought, the peaks had been cut off and planed down by waves. But the summits of the sea mounts are anywhere from half a mile to a mile or more below the surface of the sea. How they acquired their flat-topped contours is a mystery perhaps as great as that of the submarine canyons."

Standing on the bed of the sea--Unlike a typical volcanic cone. Those sounded to me like a big pile of dirt.

Actually all of the above seems to point in one direction, and that is to "construction". A steam shovel scoops out a canyon while it was above sea level and dumps the dirt in a big pile, out in the water. Then the continental shelf and the canyon was pressed down below sea level.

Then in order not to impede the ocean current the tops of the mountains that had been the ocean current's impeders were removed by the contractor.

That's really dumb isn't it? Eons ago there were no Steam Shovels and no way to cut the tops from mountains that were a mile below the surface. But suppose there were? SUPPOSE that before man existed there were "beings" able to perform construction. Suppose there were civilized, super intelligent beings in the Universe billions of years before we Homo Sapiens and our Ego believed we had evolved and formed the first civilizations.

Well I sat there out in the yard on my beach chair, I pondered, I wondered and I SUPPOSED. I contemplated, could intelligent beings a million years ago have performed construction on earth. That's silly, even stupid. I pondered about all those books that had been published about beings from outer space stopping by Earth and observing our wondrous evolutionary progress. I pondered their evidence that these visits were real. Stopping by for a visit is one thing but what my mind was saying is that they came before anything existed (In the Beginning) and they created everything by various means of "construction".

Construction would mean - language construction.
Construction would mean - botanical construction.
Construction would mean - geological construction.
Construction would mean - continental drift construction.
Construction would mean - creature construction.
Construction would mean - weather construction.
Construction would mean - environment construction
Construction would mean - the building of all things that make Earth what we know of today.

I SUPPOSED that Cloning would have been part of the "construction". The Clones produced by teams of "Beings" working in what we now call Germany would all look the same. So also would those produced by different "teams" working in what we now call Ireland look the same. That's why we still say, "He looks German" or "He looks Irish", WOW.

That reminded me of a joke about a group of Chinese people in China standing outside of an all-Chinese Hebrew Temple. One gentleman said to the other "That's the Rabbi shaking hands at the doorway". His companion looking puzzled said "That's funny, He doesn't look Jewish".

All these thoughts of construction and cloning were almost beyond my comprehension...I should have gone into the house, fixed a drink and exercised my intellect by watching TV Cartoons. I didn't. I SUPPOSED. This hypothesis would certainly clarify evolution. Continental drift would be comprehendible. Language development by teachers would make sense.

This hypothesis would make sense of everything. It would explain everything by supplying the trigger that caused the effect that explained how all these wonders could have occurred . It would be the answer great men have spent their lives searching for. "And the truth shall make men free." But would man believe this possible truth?

I got up and took a leaf from the maple tree and studied the veins and shape. WOW I thought, somebody designed this leaf on their computer or maybe right in their head then they wrote the DNA code that would build it to grow that way forever and ever.

I had read that the maple leaf has designed itself through evolution to roll up and make itself a cylinder in winds of forty miles an hour so that it would not be ripped from its tree. That makes no sense because a leaf cannot design. But WOW, "somebody" could have designed the shape of that leaf with its holding on properties. "Somebody" could...I mean somebody a real being not a supernatural spirit but a real being.

But let's not dwell on that stuff we don't believe. Lets try to do what the dictionary tells us.

Let's at least try to SUPPOSE.
"To be inclined to think, to accept as true or probable.... To assume as true for the purpose of argument"

Let's at least try to CONTEMPLATE.
"To look at attentively; hence to consider thoughtfully; meditate on; think on."

Let's try at least to PONDER.
"To be deep in thought; to think something over thoroughly."

Actually there is another problem I think you may have. I think you may be in the box. To properly SUPPOSE one must be out of the box looking in..... You have to think outside the box.

Do you SUPPOSE there might be beings from a Kingdom in Heaven who came here eons ago?

I have here proposed such a hypothesis but the final decision is up to your OPEN-MINDED thinking.

Let's try to be OPEN-MINDED.
"With an open mind, unprejudiced"

Did Ezekiel have a vision regarding Celestial Beings as religion interprets the Bible or was he actually taken up and flown to a space craft landed in the mountains. If your Jewish take up your Bible and read Ezekiel. If Christian, reach for your Douai-Rheims or King James Old Testament and make your decision after reading

Ezekiel. Is this a vision or is this his description of an actual event. What were the four living creatures he describes as having the likeness of man? Were they angels or as we say Celestial Beings or as you might phrase is were they extraterrestrials.

Do you suppose that the present East Coast of the United States once also included the continental shelf? Could you suppose that at some time eastern America was covered with water and a great canyon was created to drain it? SUPPOSE, for some reason, part of the East Coast was "pressed down" creating the continental shelf and the mysterious under the sea canyon? No, you say, you could not possibly suppose that. Well then, for now, lets just contemplate and ponder it.

Did Lot actually see and speak to a real "angel" outside of Sodom? Let's take a moment to SUPPOSE - ponder and contemplate Chapter l9 of Genesis. Chapter 19 says so very much more than its words indicate.

In Chapter 18 three "men" appear to Abraham and he honors them greatly. Two of these men departed for Sodom. Abraham then began discussions with the third man whom he called Lord about saving any righteous people in Sodom and Gomorra before they are destroyed. Abraham said "Peradventure there be fifty righteous within the city; wilt thou also destroy and not spare the place for the fifty righteous that are therein? Abraham continued his plea for the righteous in Sodom. And the Lord (an extraterrestrial?) said "If I find in Sodom fifty righteous in the city then I will spare all the place for their sake". Abraham continued his peradventuring with the Lord down to ten righteous people and the lord agreed that even for ten righteous people he would not destroy the city. The Lord then left and Abraham went about his business.

It would have been so much better for our book if Abraham had used SUPPOSE instead of Peradventure.

In Chapter 19 the two above-mentioned men arriving at Sodom are referred to as "angels" when they met with Lot. Though called Angels nothing is said about wings and there is no mention of levitation of any kind. Lot offers the angels water to wash their feet as was the custom of the time and which also indicated that they must have walked there on dirt paths and roads.

We will speak again about these Angels or as we prefer to call them "Celestial Beings"

(As information and education I have used the 1611 King James Version of the Bible to quote from. This Version was edited in 1769 to modernize spelling and punctuation. The 1610 Douai-Rheims Catholic Version was also modernized and updated. Today there are many modern versions of the Bible and you may find words like "peradventure" changed to "what if" and wearied changed to "tired". Personally I prefer the older editions and language as more poetic, rather than the revised editions that supposedly make the reading, more understandable.)

All these thoughts whirled around in our mind. Could construction be part of creation? Are some of the "men" mentioned in the Bible - extra-terrestrials? Could the area of the continental shelf have been pressed down? Would Space Aliens have an impact on the belief in Evolution? Should any of these questions be answered with a yes we might have a scientific theory right in our lap ready to be examined and horsewhipped by real scientists. WOW what a challenge!

THIS THEORY

THE SUPPOSE THEORY is not a fact. Both EVOLUTION and SUPPOSE are Imaginative thinking hypothesis of Creation drawn from Scientific Facts.

Our "Theory of Creation" is called the "THE SUPPOSE THEORY" and it requires that you use your powers of imaginative thinking in order to comprehend the hypothesis put forth by the imaginary thinking of this author. A theory is defined as set of ideas formulated (by reasoning from known facts) to explain something and a hypothesis as a supposition or conjecture put forward to account for and used as a basis for further investigation by which it may be proved or disproved. So in fact theory, hypothesis and supposition all simply mean an exercise in imaginative thinking.

All SUPPOSE requires is that the reader put aside his or her great fund of nowledge of accepted scientific theories and of religious dogma (for at least short periods of time). With this open mind we would like you to read and SUPPOSE. When you have put the book down you have our permission to resume your own scientific and religious beliefs. We are not here trying to brainwash you; we are simply looking for the truth. But until you pick up the book again (hopefully) we would like you to contemplate and ponder about what you have read.

You will note as you read, the words "our", "we" and "us" in place of my ego generated "I". This replacement is to give credit to the other part of my "I". This other part or half of me is my guarding angel, cosmic essence or spiritual being, often called Soul.

What you read here will be some thoughts that came to us as we muddled through our lifetime search for the truth. We knew that science had found "their" truth of Creation by presenting Theories as facts. We also knew that Religion had known "their" truth of Creation for thousands of years without one iota of doubt. But we think we may have come upon a truth of Creation that is logical, reasonable, easily understood, has a cause and effect element is based on scientific fact and can actually answer the questions that have bewildered mankind for all the centuries of our existence.

The things we have to explain in this book are truly controversial; they are so beyond the normal "acceptable" scientific and religious dogmas that we are most apprehensive about our ability to make our thoughts clear.

Let us speak again now about Genesis for a few minutes. We know that no one was there "In the Beginning" except for God, of course. This then would seem to makes Genesis a pretty good factual record of how things were at that time. We must though keep in mind the many, many, times God's record was rewritten, translated, and "improved" by many hundreds of well meaning scribes, and also many religions fanatics, over the six or ten thousand years of it's existence. So lets just keep that in mind.

Since this descriptive part of Genesis does not deal with Morality we feel it is far more likely to be close to its original composition. The record that God gave us lets us know that in the beginning the Earth was, without form and void and darkness was upon the face of the deep. God then said, let there be light, and there was light. He then divided the light from the darkness. That sounds to us to be a lot more work than could be completed in one day, so we think he spoke in days instead of centuries and in a manner that would be acceptable and miraculous to the superstitious people of that time

The darkness upon the face of the deep would seem to indicate that the Earth was not rotating, in the beginning. We believe we would all have to conclude that the Sun was there at that time and was the source of the light. If then the darkness were always upon the face of the deep, at that time, it would have had to be facing outer space away from the Sun on that side. The other side of our Planet would have had to be always facing the heat of the Sun; thusly it could not have been rotating, at that time.

We believe then that Genesis tells us that one side of our Planet was solidified, it must follow then that the other side would have had to be still molten from the heat of the Sun. Then we wondered what science had to say on that subject and HALLELUIAH.

Science "agrees" with Religion and like God, who named the light Day and the darkness Night, science named the solidified super continent side PANGEA. Pangea was later split into the Northern Hemisphere, called Laurasia and the Southern Hemisphere called Gondwana. Neither science nor religion has much to say about the molten part of our Imaginative Thinking. Genesis says "And the Spirit of God moved upon the face of the waters", which certainly doesn't leave much room for Molten. Science doesn't seem to address the Molten, they seem content in believing that some sort of consistent under-the-Crust current of Magma pushed Laurasia and Gondwana in a multitude of directions, like North, South, East, West and a variety combinations of all four. This was all done on top of Granite made solid by all the water covering the Blue Planet. They also seem to believe the first Oceans where Red.

What about the Rotating part, you may ask, and our answer is that the first question that will come up is how the blazes could the Earth "Begin to Rotate". In answer, Religion will state the unquestionable Supernatural Powers of God. Science will invent a comet of gigantic proportions that whizzed by Earth so close it sent us spinning and then it continued its journey into deep outer space never to be seen of again. The question about Rotation that you may have asked has another answer, but it deals with a, shall we say, technology, of such gargantuan proportions, it will take considerable explanations. Since we have not yet discussed this technology we will leave that answer for another time.

Oh, by the way, as you know, Earth is the Third Planet from the Sun. Our Imaginative Thinking does not believe Third Base was our position - In the Beginning. By combining our thinking and our Reasoning, (which will be called Idiocy) we have concluded that the Earth (after the Big Bang) ended up between the Sun and Mercury. It was able thereby to maintain it's one-side-molten status, as it would not have been hot enough where we are now for that kind of one-side-molten maintenance. We realize you had a difficult time over getting the Earth to rotate, now we have to move Earth Two Planets further from the Sun. I'm sure that neither Religion nor Science would address this hypothesis, even if they could stop laughing. As soon as we discuss that technology we spoke of we can then answer both of these dilemmas.

Speaking of that technology we spoke of, let's look at ours. We have Radio, Television, Sonar, Radar; we've got those there airy-o-planes, submarines, rockets to outer space, atomic and hydrogen bombs, the Stealth Bomber and even a space station. Another acquisition to our accomplishments is the very secret AREA 51. We truly believe that in the not too distant future, we will develop an anti gravity machine. So also will we develop the ability to produce a force field with Electro Magnetism, and many more technical advances. These advances will be use in our exploration of space and the establishment of bases on the Moon and on Mars.

We may suggest that your concept of intelligent living beings from a location other than Earth might need to be "entertained", and possibly altered. So rather than the inappropriate words--Extraterrestrial or Aliens our reference will be to "Celestial Beings", "The Creators" or even "The Gods".

We have always considered SUPPOSE to be a non fiction, science fiction piece of work and we are quite sure that you will consider it a piece of work, also. Upon exposing friends to my hypothetical supposition that beings from a Kingdom in Heaven landed here in Pre-historic times, and engaged in construction they ask, "Why would they come here?" "What were they looking for?" "Why would they want to construct by creating creatures and plants here? And other questions about their motives.

In our endeavor to clarify the motives, morals and technology of our celestial beings, we have created this science fiction section to answer your question of why they came here. This section also deals with and explains their Gargantuan Technology we spoke of above. "They" of course are what you have always thought of as a joke. "They" are the extraterrestrials of SUPPOSE. "They" are the creators.

A speck - We suppose you could have called it, a speck, when observed from space, even though it was at least 15 miles in diameter The speck was a craft carrying intelligent life into a newly discovered solar system. The craft was from a Great Kingdom in Heaven and had for many decades been traveling the various Universes in space searching for the right place to conduct "Experiments". The hieroglyphics imprinted on the craft translated to the meaning “The Searcher”.

The Searcher had entered many other Solar Systems in its quest but had not as yet found the perfect laboratory. We do not mean they were looking to find a ready-made system to work in but rather a Planet that met the criteria that the Council had set forth. The Planet and Solar System, for which they searched, had to contain no intelligent or any established life forms, it had to be separated from their Kingdom by a prescribed number of Light Years. The Planet they searched for had to have the potential for the development of a great variety of climatic and atmospheric conditions.

The Planets, so far found, in their search, had not been of acceptable conditions as most had solidified from their original molten state or were still molten, ablaze with fire or enveloped in various forms of gas. The Council had ruled that the acceptable Planet had to have the potential to produce at least twenty separate, water isolated land units on which various types of experiments could be conducted. As each unit would be self-contained no experiment could spread over the entire Planet, until desired to do so.

Commander Joseph Hovar’s hopes were, as always, high as he guided his craft towards the star that gave energy to the System. He studied the computer screen as the reports began coming in from the unmanned observer probes that had been sent to each of the Planets. Several contained water, a few had an atmosphere of chemical substances that would need to be analyzed, and possibly collected and stored aboard if determined that they might be of some use in a future time.

There was also a great air of excitement among the ten thousand Celestial Beings on the craft. They also watched the ships communicating receivers as the reports came in. They did not see what they hoped for. They did not see the planet that would fit the criteria of their quest.

These Celestial Beings were all Scientists - groups of highly educated experts of every scientific field in the Universe. Highly educated experts with absolutely no experience. The Council, which ruled their Kingdom, had hundreds of years ago forbidden any and all scientific endeavors except 'study". There was, of course, good logical and reasonable cause for the Council to have banned any physical experiments by science.

Had the Councils of many centuries before exhibited even a little control over its scientists the Kingdom might not have suffered the "Great Failure", might not have almost ceased to exist. That failure was now ancient history, ancient history the "Searcher" existed to make just an incident of the past.

Back in those ancient times, science ruled. Everything in the Kingdom that was good, convenient, enjoyable, labor saving was supplied by science because everyone was some types of a scientist. It was good- and then the new science of Genetic Engineering was developed.

Genetic Engineering was also good for the Kingdom - vegetation was improved, size was increased, flavor of food enhanced, color was altered, even taste changed, all with Genetic Engineering. From vegetation it evolved to the creatures used by the inhabitants as food nourishment. Year after year more things were Genetically altered, to improve them - It was good. Since it was good the Council never considered instituting any restraints and Ethical issues were left to each experimenter to consider. The actual "Evolution" of Engineering of Genetics advanced, of course, to the Celestial Beings themselves. It was good. Small changes made hereditary diseases disappear - Fat cells were eliminated from birth - obesity disappeared. After a few generations nametags were required, as everyone in each family looked the same.

It was a perfect world, an idealistic existence, no problems, and no sickness, just utter perfection. The council did meet on occasion to vote on some financial matter or to change the seating arrangement, because of the reapportioning of member's districts. It was at one of these meetings when the traditional (and symbolic) call for new, or problematic business was announced, that the representative of one of the outer planets raised his hand and stood to address the Council. He informed the members that newly planted seeds had not grown more than four inches in height then they drooped and died, in field after field. New seeds fresh from the Genetic Factory were rushed to replace the defective ones and they too refused to grow. He requested of the Council that they appropriate funds for the shipment of food to his planet, as their budget could not support that additional expense. Most of their farms had been planted twice and in some cases three times. The growing season would not permit further plantings.

Within a week report after report arrived at Council Headquarters, new vegetation seeds would not grow. So much had over the past few generations, been added to improve vegetation and so much had been removed that their ability to grow was gone. Then the most devastating reports began arriving at Council Headquarters... Malnutrition... Most all the Planets of the Kingdom began observing the effects of Malnutrition on its citizens. The food they had been eating had increasingly been decreasing in nourishment The Council reestablished the Food Testing Bureau and their tests proved that most food did not have enough nourishment in them to sustain any kind of life.

Scientists stuttered and stammered...but the color of food was perfect, the skin never got bruised in shipping, its aroma was delicious, it stayed fresh for months, its taste was better than ever before in history. Don't worry, we'll fix it. And they went to work.

They worked and they worked and they worked and they couldn't do it.

Then came the straw that broke the camel's back. WATER ...Water, by the wonder of science had been artificially manufactured in Oxygen-Hydrogen Factories built at great expense on the arid planet of the now royal and ruling class and gave them all the water they could use and more. Water was manufactured and then stored in great constructed fresh water lakes. The procedure for producing water in these plants was a very noisy one. The noise annoyed the Royals and the Council.

At monumental expense the water factories were demolished and rebuilt on a small sparsely populated planet within their kingdom in heaven. Very large space ship tank crafts were built to supply these great lakes with their normal level of fresh water. The tank ships shuttled silently back and forth between the water factory plant and the giant landing field at the location of the original water factory.

Again and now with water, sloppy science showed its head. The original Water Factor was on an arid planet, which had tropical weather temperatures. The planet to which the factory was moved was one of fog and chilly weather. The tank spacecrafts by working around the clock managed very rapidly to bring the water level of the great lakes back up to normal. The level had dropped during the relocation of the factory. Science made no changes in the machinery or procedures when the location of the plants changed. It was a matter of bringing up the water level quick.

The lake water level did come up quickly... BUT... due to the changes in atmospheric conditions the water coming out of the factory was SALT WATER. Science was so sure of its own genius and infallibility that no tests were made of the water before it was transported.

After two hundred years of applying the wonders of science and Genetic Engineering, all Science became known as the science of "UNNATURAL ALTERATION", as the ability of vegetation seeds to grow was gone and the fresh water supply was contaminated. Science said they could fix it. They tried ... they could not.

Oh my gosh, you don't possibly SUPPOSE that our Earth Scientists diligently using their new tool of Genetic Engineering by altering our fruits and vegetables, cloning our cattle, chickens and other farm animals might be causing Earth's Mankind problems in the future. Even though there isn't at this time an Associated Scientific Lobby for Ethical Engineering Practices (ASLEEP), I'm sure each Genetic Scientist considers all side effects that might occur from his experiments with the unknown. Science would of course consider the effect of allowing any of their cloning experiments to mate with other creatures they consider to have evolved by nature. Such mingling could have dire consequences.

Food sources other than vegetation also suffered a set back famine ensued death by starvation "became common place". The food supply was bad enough to suffer but then reports of mutations in their own species began to multiply

The once magnificent science of "Genetic Engineering" brought adoration and praise to scientists throughout the Kingdom for over two hundred years. Now within a year, the science of unnatural alteration brought scorn and hate down upon those unethical - unthinking - despicable scientifically moronic geniuses.

It did not take the Council long to re-assume its authority and begin handling the job of being rulers. The first edicts regarded, as should be, their own species. Any families producing mutated offspring were made sterile to stop the mutation. The Mutants were destroyed. The race had to be saved. Artificial food was chemically produced and slowed the figures of death by starvation.

Though it might seem self defeating one of the edicts declared by the council was that no, absolutely no, scientific or Genetic experiments were to be carried out anywhere within the Kingdom. Science had caused enough destruction.

Since we know the Searcher had aboard ten thousand beings we know their race did not die out. Let us tell you how that came about. For one thing the knowledge of how to produce artificial food came from very ancient processes once used in their very first space expeditions. Also the tank craft shuttles now had to make long and dangerous trips to planets outside the Kingdom to bring in fresh water since the Council would not approve any "scientific" repairs to the Water Factory.

These Beings from a Kingdom in Far Outer Space are not Homo Sapiens, by any means. Let us try here without going into to much detail give a very basic explanation of their reproductive procedure. Their intellect - intelligence - gained knowledge - memory - character - their entire consciousness is contained in a Spiritual Essence that lives forever and ever - provided of course it has a biological body in which to reside. Each individual reproduces bodies in a solo manner at prescribed intervals of time. It is a very difficult procedure as the new body is formed within an egg like pod and must be carried until the new body is formed and basic knowledge of how to function is absorbed. At this point the Pod emerges and is opened revealing a miniature child like, more or less, clone of the producing being. The new creature becomes a companion and friend of its producer while it grows to maturity. It also becomes what to us might seem a lover. As part of the maturing the two bodies engage in activities that produce a similar exhilaration as that which we would experience through sexual activity. During each of these matings, portions of the Spiritual Essence pass to the new biological body. As the producer drains his form of spirit he begins to die. The new form, containing the old knowledge, cares for its "father" and when deceased -- see to it with great pomp and ceremony that it is laid to rest in peace and with great dignity.

Over the Centuries accidental deaths did occur. Explosions - fire - crashes - drownings - falls all caused biological bodies to die. Since the Spiritual knowledge had to be saved, there was a "surrogate class" of being whose only function was to produce new biological beings. The bodies they produced were of blank minds and could absorb and hold for transferal, the Essence of an accidental death being.

With this very basic information in mind we can now continue with how the doomed race survived.

There were many dissidents; actually they were early anti-unnatural-alteration proponents (considered weirdoes) in those ancient times that caused much disruption and actions that were deemed unacceptable by the Council. Since they would not join in the advances of science and simply existed to slow progress. They were classed as troublemakers and shipped to colonies in the outer reaches of the Kingdom along with a great number of the Surrogate Class, which became obsolete, because of the advances in Genetics. The dissidents used no genetically altered elements and were able to supply original seed to the Council in this time of most need.

During the pre-recorded ancient history of the Kingdom only war between planets existed. The only types of beings that existed at that time were warriors. As time and struggle advanced peace developed and the cruel and aggressive warriors that could not accept peace were executed and their Spiritual Essence was also sent to the outer planets for storage.

We hope you can now understand why there was such crackling of excitement and anticipation throughout this craft full of scientific geniuses. If they find the authorized planet they can set up land laboratories and actually conduct experiments. In this way their Kingdom in Heaven would be in no danger.

The Planet of most interest to Commander Hovar was the one closest to this Solar Systems Energy Star. A probe had not been sent to this Planet because of the Commanders personal interest in it and his hand actually trembled as he turned the Searchers Scope towards it and moved in closer.

The first Planet from the Sun did not revolve. Because of the heat of the Sun, the side facing that Star had never solidified as had all the other Planets in this system. The side facing the Heavens was a great plain of solidified granite. You might say it was without form and void and with no light source from space there was darkness on its surface.

Everyone on the Searcher was elated as the Commanders voice crackled through the ship "This looks like it". Probes indicated traces of water near the areas where the Granite and the molten matter met. When finally the landing was made and samples taken the joy of finding one cell life was beyond their greatest expectations. With this original planet life, they could engineer to their hearts content and never a worry about endangering their own kind again.

If only Darwin could have been there when they first placed a one cell asexual creature under their microscope and programmed it to split and produce daughter cells. Oh how he would have danced and jumped with joy with them as they joined several together and designed a muscle system that gave it actual locomotion. He would have run from one lab to another watching the group that was designing the digestive tract. Then rushing across the corridor to the optic lab and been awed by the designing of eyes. These eyes would later be connected to a brain by nerves, being designed by another team. Darwin would though have been surprised that a complete segmented worm could be designed and created in less than a month. He'd wander around saying "You mean it doesn't always take a long - long time."

The Commander could not have asked for more--He always "strove for perfection" and everything pointed to this Solar System and this Planet he hovered over as being the "Perfect Answer" to the Councils requirements. He had found Perfection.

"Scout out a landing sight", he ordered and after a few passes over what would later be called the Northern Hemisphere of this Planet he settled the humongous ship to rest on the cold void granite surface. And so we witness here the first landing of a craft on the Planet EARTH.

An Electromagnetic force field was set up immediately to protect the metropolis size ship and its disembarking passengers from the millions of large and small asteroids that flew through the atmosphere and crashed into the surface at regular intervals. It was evident that most of the Planet had received hits by asteroids from the number of craters caused by the collisions as they and after they had solidified from their molten state. This planet also showed evidence of many very large strikes.

Since Commander Hovar intended to build a City Laboratory for over 6,000 of his people who would start work here, while he returned to report to the Council. He had to insure their safety. The Force Field gave good temporary protection for their Space Station but for the long run the Asteroid Problem had to be solved. Every work ship aboard was sent out to capture every piece of the millions of asteroid space junk between the Sun and Jupiter. This space junk was then to be placed in a ring or belt that would orbit between Mars and Jupiter. This location was decided on by the Commander and his staff as the most reasonable location, falling within the gravitational pull of Jupiter. Any Asteroids breaking away from its belt due to a collision with a fellow traveler would be pulled into Jupiter, which could handle any strike, there-by insuring the Station and the other smaller planets from catastrophic collisions. It was a tedious and time consuming project as the ships had to grasp the asteroids and place them within the designated area prescribed for the "Belts" orbit. Systems were set up whereby asteroids and comets from outer space would be tracked and if their course would cause trouble to Earth they would be diverted months before visible.

As this procedure continued, material was unloaded from the Searcher under the protection of its Force Field and the first buildings began to be erected on the surface. They had artificially mimicked the energy strength of what we would later call a "Black Hole". This energy force would need to be installed in a Generating Building to supply all the energy required by this Surface City after the Searcher had left to return home.

On the surface, Electromagnetic Forces were employed in the unloading of the buildings and materials, along with Anti Gravity machines, which were all part of their common every day equipment much as a Fork Lift Truck is to us today. One Celestial Being holding a small box with hand controls was able to move extremely large pieces weighing tons, with no effort and with no noise.

All this equipment had been carried so long and never used before, as a surface station had never before been required. Tons of the same Artificial Food used to fight the Kingdoms past tragedy, were unloaded along with heavy generators, bulky buildings walls, ceilings, floors, furniture and all sorts of laboratory equipment.

As Heavenly and almost Holy their name of Celestial Beings, might sound, please keep in mind that they have body waste and their ships toilet waste holding tanks needed to be emptied from the Searcher before it left for home.

They did consider their body waste and food garbage as contamination and they were obligated not to contaminate any part of any Solar System. So using a laser or pulverizing type beam, they cut channels or vein-like shafts into the surface of the Planet and pumped their liquefied waste from the holding tanks into the vein cuts. To complete the disposal process great pressure and heat was applied aiding the chemicals that had been added in forcing the water out and solidifying the waste to a black rock we would later call coal. As other Star Ships were called in to participate in Project Earth they too cut veins into their landing areas in the Northern Hemisphere and emptied their holding tanks into the deepest part of the veins. This almost pure waste, which is to say they had no fossil ingredient in them, would later be labeled Anthracite.

The searcher was not the only Star Ship that had been dispatched to satisfy the Councils needs. Six more were investigating other areas of the various Universes. Hovar sent both electronic and telepathy messages to all of them that they were needed at the Milky Way as soon as possible. Their goal had been found.

As soon as the first four arrived they and the Searcher began the process of pushing and pulling the Earth further back from the Sun to occupy its space with the Planet Moon. The Celestial Beings on the surface at that time had to get back aboard the Searcher. The type of power employed, which mimicked that of a Black Hole would be much too strong for their survival. As soon as our Planet was in place and its relationship with the Moon established it was manipulated in such a way to cause it to revolve. After the move was accomplished the six thousand were returned to the surface to begin their work and Hovar set off for the Kingdom with samples, pictures and enthusiasm. The Council was overjoyed. Hovar, then as the Commander of Earth Project, soon returned.

Now that was an imaginative thinking Science Fiction story. Its purpose was to "suggest" their possible motives, "suggest" their incredible technology and a little insight into their morality. Speaking of them, Celestial Beings, we would like you to know that they and their flying saucers are the basis of the Suppose theory.

Oh Boy.... That statement just alienated and possibly drove away about 30% of the readers. You just have to try to be more open-minded.

Oh well ...there are two other theories floating around. One is the Religious Creation theory and our logic and reason is inclined to believe that that one is based on superstition and magic and would you believe I just lost another 30% of my readers.

Well Gee.... That leaves me with 40% of my readers. The other theory of creation, as you know is the childish daydreaming of Darwin. His experiment with imaginative thinking didn't turn out to well. If any of you 40%ers have ever applied just a tiny bit of either your logic or your common sense you would have concluded that there is absolutely no way Darwin's thinking could be correct. He thought that our nervous-visual, digestive, hearing and blood systems, with their complexities all working together as an almost perfect unit had managed to evolve by some kind of a haphazard uncontrolled and seemingly magical manner. Of course you concluded that he was wrong, since no other conclusion makes any sense at all... and there goes another 30% of my readers.

That's bad news... but the good news is that we now have 10% of the most highly intelligent, open minded logical, reasonable minds to continue in this quest for understanding, this search for truth and that is what SUPPOSE is.

The problem of understanding is that there are a lot of little facts that are misrepresented by our scientific and religious scholars. Things like the fossil records that prove evolution. Actually these fossil skeleton records do not show the progression of a natural process of Evolution by nature in a so-called natural manner. These records do show the "evolution" of the technological development of Genetic Engineering by natural super beings. (As opposed to a super-natural being). These records also demonstrate the many design changes "created" by that technology.

Our religious scholars did not make it clear that when, whom we call God, separated the Day from the Night and found it to be good. He did so by causing the Earth to rotate rather than just to cause the sun to rise in the East and set in the West. Thereby revolving around the Earth, as they once believed it did.

Our great men of science accepted this misconception of an ever-rotating Earth even as they argued the unscientific idea of the church that the Sun revolves around Earth.

The reason we of SUPPOSE are so smart is that we read in Genesis that the Earth was without form and void and darkness was upon the face of the deep. Then we read of Geological Studies that indicated only one landmass, in the beginning. We reasoned that if there was one landmass and it was dark then the Earth was not revolving and one side, towards space was cold and dark and had solidified. The other side faced the sun and was still molten. Since it was still molten it was closer to the Sun than Mercury is today. And yes, one of the reasons for the Science Fiction Story was to try to get you to think Scientific Fiction, for we believe that Celestial Beings, eons and eons ago possessed technology we can now only conceive as Science Fiction. We believe that with this technology the Earth was moved from this too hot for life position, passed Mercury and Mars and put into orbit with the planet Moon.

The uniqueness of our orbit should give us food for thought. In effect there is a center of gravity that orbits the sun while Earth and Moon orbit that center of gravity in a kind of a tumbling action.

In our imaginative thinking story we spoke of the asteroid belt in orbit between Mars and Jupiter. Science should have said "Holy Crap" it looks like someone collected all this asteroid space junk and put it in a ring orbiting the Sun. Instead they made up stories about maybe a planet blew up and then magically spread out in a ring around the Sun. (They would not have said Magically). As far as the asteroid belt is concerned we believe that celestial beings collected all that asteroid space junk and put it in a ring orbiting the Sun.

We also believe that we Earthlings are going to have to collect all the space junk we have left up there. Put it in a ring around Earth before it causes the death of an astronaut or the destruction of a space station or craft. And that is not Science Fiction.

But as we know how we humans think and work. We are sure that no government and no organization are apt to appropriate any funds for cleaning up after ourselves until there has been a catastrophic incident and the loss of multiple lives.

Negligence in keeping the space around earth clean of debris might well result in a horrendous mass murder and that would be an abomination. But there is also joy in some mass murder and that is what I wish to participate in. Put on a Hospital Operating Room Gown if you have one and join with me. Together we will now, with a mass of joy, murder Evolution.

EVOLUTION

Before we start we wish to express a very truthful truth. Evolution is a theory. Evolution is not a fact.

Scientists have also stated that NATURE can accomplish deeds of evolution over extended periods of time. Students have been taught this fallacy also as a fact, it is not a fact. NATURE never has nor never will accomplish any kind of scientific deeds. (Unless one believes in magic.)

I hope that the following attempt to murder evolution will be as enjoyable reading to the believers as the writing of it is to this disbeliever. I accept your sympathy for my uneducated dalliance into subjects that are way over my head. Never the less I ask that you SUPPOSE, that you ponder, and that you use "some" common sense in your contemplation. I humbly thank you.

How well the Professors have gotten across their teachings. We must congratulate them. We recently spoke to a college educated person, suggesting that we did not concur with the belief that man had evolved. We were met with a strong resistance to any such hypocrisy. "Of course we evolved, cells came together, and we evolved over a long period of time." We believed, because of her religious background that she might not have agreed with us regarding evolution but would have insisted on the truth of Biblical Creation. It was however Evolution that strongly won.

We have not been the first to realize that Darwin was mistaken in his imaginative thinking, even though it is obvious. But "the grip" Darwin's Theory has on the minds of all the educated students of the world is mind boggling. We believe this grip is so strong because much of Darwin's imaginative thinking hypotheses are taught as fact. After all what student would not believe in facts. There are so many charts, diagrams, books and movies, rationalizing how evolution actually happened, that assail a student looking to be educated, they can hardly be faulted for opening up their minds and absorbing the "factual" fallacy of evolution. In truth all this literature on evolution is actually literature on the Fossil Record. Everything tells them this is the truth, learn it. And they do. Actually they have to learn it by heart because the questions on the test are based on the garbage in their books. Then again there is the social life of college and the partying. One doesn't have time to think about it, just barely time to study and learn it. So thus the reasons for "the grip" After all it is by this system of memorizing those things, which are the accepted things that culminate into that thing which is called education. Students cannot be educated on the facts of evolution, because there are none. Just try to find out how the coming together of single cells happened, just try to find out how all these daughter cells could create living creatures without creating. Darwin thought that there were ways by the laws of physics that things just happened over long periods of time and that these happenings just seemed like design. It looks like design, it smells like design, it acts like design but is just an illusion of design.

Not every one can remember 70 years ago, but I do, I was 12 then. I remember standing before the evolution wall in the Museum of Natural History, in New York. It was a glass-enclosed floor to ceiling display of the fossil record of human evolution. How could anyone deny evolution, there it was right on the wall. As far as I was concerned those were the actual skeletons. There was proof on the wall. I strongly believed. I was 12.

I guess that like those college students I was a little to close to the wall as they were to their books. When later I was able to step back and give it some thought the picture changed some. "Contemplation" made the evolutionary idea of haphazard events seem very orderly. Science tells us the one cell creatures from which we came managed somehow (I know it was not by magic, but somehow) to join together and then reproduce themselves. All this was by chance. Even though this joining together was a random event it was done in an orderly manner. The cells first came together and created, oh sorry there's no creation in evolution; formed things of soft body like segmented worms and jellyfish. Almost as if they somehow figured it out they started to come together with hard shells or armor on the outside to protect their soft insides.

We can't say their intellect had helped them even though by now they had haphazardly developed brains. During all this evolutionary development they managed to get themselves a digestive system and a system of nerves which included optics. Then by this evolutionary power of theirs they moved the outside shell into the inside by designing, Oh, Sorry there's no design in evolution, it's just an illusion of design. I mean forming this well constructed backbone. Since their protective shell was now inside they managed to evolutionize themselves scales to protect their soft sides. "Just imagine" from the first two cells that bumped into one another by accident and joined together MAN developed. If it weren't so very scientific it would be magic.

Actually one would have to "just imagine it" since it has no "logic" to sustain it and no "reason" to perpetuate its rationalization. One might wonder how something imagined, could exist for so long and have the support of such highly learned men. Having already done that wondering we have concluded that the learned men not only taught evolution as fact but also accepted it as fact, themselves.

I must suppose that if evolution had consisted of, lets say, three or four one-cell entities coming together and forming a segmented worm that had some locomotion we might have believed that. We even might have considered it an illusion of design. However once this random collision of cells somehow developed a mouth and a digestive system then that would be just to illogical for any intelligent mind to accept. Mr. Darwin did not dwell on this beginning of evolution or how these worms had managed to reproduce themselves in order that they should survive as the fittest. He just seemed to accept that in the beginning the jellyfish and worms began having sex and propagating their species to become the fittest.

The science of evolution mostly deals with the higher class of development. Once it got started (Somehow) it just kept on evolving right up to mankind. Cells just came accidentally together (Somehow) and in a random haphazardness ended up with a hit and miss procedure that produced (Somehow) the many complex systems that we possess. What a wonder that we can see, and in color (Somehow). It would almost seem that these cells had a collective intelligence since they managed (Somehow) to keep improving. They did manage to bring forth a brain, a nervous and digestive system

(Somehow) thereby producing two almost perfect homo-sapiens creatures called Man and Woman (Somehow) We must also mention the blood system and the amazing and wondrous somehow sex and reproduction systems.

Instead of the Theory of Evolution maybe we should call it the Theory of Somehow.

I don't mean to be disrespectful or sarcastic towards science, I just can't help it. The thing about this haphazard coming together of cells that truly awes me is that all of our produced by hit and miss bodily functions joined up in such a way that they seemed to support each other. Almost as though they were designed to work together. But of coarse all educated people know that what appears to be design is unquestionably and simply randomness in the coming together of cells.

All educated people know that Evolution is the random coming together of cells in such a way as to produce all of Earths living creatures and all of our botanical specimens without design or creation ever being involved.

Although I was never an educated person I was for some time an Evolutionist. It is not our endeavor to here alienate any readers of either of those classes with our paternal sarcasm. The endeavor we have in mind is always to try to get the reader to suppose, to contemplate, and to ponder. So please don't let me build any walls of discontent. My Doctor has instructed me to stay away from salt, but I'm sure you can take what I say with a little of it. Also keep in mind that before I retired I was nervous and jerky but I'm no longer nervous.

I do not know if students of evolution are cognizant of problems in evolution. They must, of course, but I imagine that they are told that that is the way of nature. I read about those first fish that were not happy by being pushed around by ocean currents and looked above the surface of the waves and saw Coney Island Beach or Malibu or one of earths many other beaches. Having previously taught themselves to extract air from water through their gills they decided to learn how to breathe pure air. This was done with the new lungs their cells had mysteriously and suddenly developed. It then became necessary to get out of the water or drown. They persevered and in a short time turned their fins into legs and walked up on the beach. I do not know how much time this process must have taken but it was necessary for their survival. Then after another long hard struggle the cells of these fish turned them into mammals. The problem I have is that some of these mammals then looked at the water, turned their legs back into fins and keeping their lungs returned to the water. I guess that's just nature.

When I was a little kid I loved the giraffes at the Bronx Zoo. In school it was explained to me that by reaching higher and higher up a tree for food, their necks grew longer and longer. (This was confirmed later in the Origin of the Species.) Even as an innocent kid I pondered that one. Why didn't the hungry cow under the other tree stretch her neck for food? What about the gazelle they must have eaten from the same tree. They told me that was just how nature worked.

... I do not believe there is anything natural in nature...

How dreary to bitch so much on evolution but what are the answers to all these problems. There are answers that remove the differences between your thinking and my thinking. The answers create another problem...That you won't believe them...So again I ask for your open-minded pondering, your contemplation abilities and your willingness to SUPPOSE.

As you know we have proposed that as Genesis put it "In the Beginning" beings from a Kingdom in Heaven arrived on Earth and began many eons of development on this planet. What science named Evolution was one of their design developments. Actually Evolution is the record of the systematic advancement of their Genetic Engineering technology. Advancements in design through the ages as they developed various species from the one-cell beings they found here. In laboratories they joined together groups of cells and using DNA Codes, programmed these cells and built digestive, muscular and optic-nerve systems. They designed and created Jellyfish and other soft beings. They discovered that by using calcium and other chemicals they found here they could develop a hard shell to incase their soft-bodied creatures. Working with the one-cell creatures they found here was a unique experience for them and they relished the designing and the creation that they could and did accomplish.

By designing a backbone and scales they came up with a fish that used its newly designed gills to breathe. I'm sure the creators where jubilant. As no creature ever adjusted to its environment the creators took the fish into their laboratory and genetically engineered it for life on land when that environment was available. They designed lungs and engineered the fins into legs. Many trial and error experiments where conducted. These first land fish still laying eggs where probably froglike creatures but only existed in the laboratory having no presents in the fossil record. Continued experiments with land animal reproduction developed the mammal. When the environment developed sufficiently to support the first mammals they were released into it. Never did any creature adjust to an environment. The environment was always prepared first and then the creature introduced into it.

As far as the giraffe is concerned here is how that came about. Be it known that there were thousands of teams of celestial beings working on creation. The team creating the giraffe was experimenting with blood flow and nerve transmission to the brain. They were also studying the effect a long neck would have on the esophagus. In preparation for that experiment the neck of the giraffe was genetically engineered to grow to its astounding length. We must also consider that there had been dinosaurs with very long necks. The dinosaurs were long gone but they still had the DNA plans on file. They never even thought of an apple way up on a tree. That was another of Darwin's erroneous conclusions. They demonstrated that even if a neck is long the blood and nerve system will work as intended and they made adjustments to accomplish that. As far as that cow under the other tree, those creators where solely concerned with a sack for the milk and the size, shape and number of teats in the cows groan area.

As we are here wondering, so also did the creators. They wondered if this newly created land animal the mammal could live in the water. They then designed a fish like creature mammal and took it from their laboratory, probably in Australia to where they had constructed fish pens with coral. We call the pens a barrier reef, although they have now all been open. By confining these mammals within the barrier pens they could more easily be observed and necessary adjustments made.

I'm asking that you SUPPOSE in regard to this new way of thinking. The design improvements of all living things as illustrated by our fossil records obviously demonstrate an advance in genetic engineering technology. Such intricate design changes in bone and organic structures had to be "accomplished" and could never have just happened...haphazardly.

What Darwin did was to look at the fossil record and ponder. "How the Hell could all these complex design changes have occurred without any control?". He listened but no answer came back to him. As a scientist of that period he absolutely rejected the religious dogma of Creation. "How the Hell", he thought. "Well I guess it must have "just happened" I guess that's just the way nature is I guess. I know it's not logical or reasonable but if that other guy gets his similar theory out before me, he'll get the credit. Well I want the credit and I'm going to put out my theory first and get the credit. The survival of the sexiest might sound dirty so I'll call it the Survival of the Fittest and say it all just happened".

So we got the "Origin of the Species" in the middle of the eighteen hundreds. A million scientists didn't scream "That's stupid" "It couldn't have just happened" "It doesn't make any sense" "It's not even logical" "That's ridiculous". In the middle of the nineteen hundreds it was still taught as truth because they still had no answer to how the Hell it all could have happened.

We are now in the twenty-first century and our intellect is still chained by nineteen-century thinking. Thinking that is based on an inability to comprehend how the design changes of the fossil record could have happened. This inability to comprehend has created the theory of evolution.

By the way, we did go on and on knocking education, we're sorry. It is only evolution that thoroughly needed to be knocked. The silliness of evolution should have kept it out of the educational system.

So you see I am not an Ogre out to shut down all the universities of worldly civilization and culture

History has its dates to remember,
Poetry it's lines.

Memorizing is very good
For all our students minds

But don't remember silly stuff,
Say, wait-hold-wait enough is enough

Refuse the course on evolution,
Its teachings are absolutely no solution.

Choose the worlds great literature,
To make you wise and to mature.

And I also now propose,
That you read the book, SUPPOSE.

Can you believe that there are some readers out there who do not believe that I came even close to murdering evolution? After all they lament the fossil record proves evolution. The fossil record--The fossil record--Oh such faith in the fossil record. Take those "bones" away from Darwin and his theory is just a lot of gibberish. The theory of evolution and Darwin's "illusion of design" cannot stand-alone. Without your human bones you would crumble. Without fossil bones imaginative evolution too would collapse. Help me with the slaying of this evil thing. With an open mind take another look at evolution. Look with new eyes that have never gazed at that bony record. Look with a clear mind that has never studied the changes in design of various creatures. Let open mindedness mean new eyes and clear mind...then see if you can believe the haphazardness and randomness of "that" theory.

Evolution depends on the crutches of bone to hold itself up. Evolution depends on the truth of the Fossil Record to give it body, to give it life. Without fossil bones and on its own evolution is unsustainable.

Now look at our SUPPOSE Theory
Its heavenly beings may look eerie.

The idea of it may be hard to take.
But it's not an intellectual fake.

You have only to believe in "them"
The biblical angels that "appear" to be men

Soon you'd come to see their deeds
As the hands that creation needs.

They triggered the one-cell with their D N A,
And so life began - what more can we say.

They cloned and made D N A Code changes galore,
And produced all the creatures one could ever wish for.

Poor Darwin needs crutches of Fossil bones
And gives no credit to creations clones.

But SUPPOSE can stand staunchly aloof and alone.
Never needing the support of one little bone.

One more thought we think you should entertain. This question has been put to me. "Who the Hell do you think you are to stand up and contradict thousands of years of religious belief and hundreds of years of scientific knowledge? We have pondered this question and have decided that it is not "Who" we think we are but rather "Why" have we done it! The because of that inquiry is that here in these thoughts, here in these words one can find TRUTHS that even if not sort, must be contemplated.

That contemplation is for you to do.

I truthfully do not mind being called a fool and an idiot. I do not have any scholarly letters of any kind to put in front of or behind my name. The best I can manage is "High School Graduate" (HSG). Even with this unlearned ness dragging me under I can say with humility and pride "Science is wrong if it believes that the principles of evolution are to be credited with the creation of life and growth on our planet". Creation is a humbling and awe inspiring circumstance. All credit for creation goes to and only to God the Creator.

CREATION

Creation was a slow and complex procedure.

We should be humbled by its Magnificence.

We should be perplexed by its complexity.

We should be astounded by its diversity.

We should be awed by its beauty.

We should be engulfed by its Glory.

We should be bedazzled by its wonder.

But more than just Adjectives we MUST praise the Creators for their deeds.
We must thank them for their dedication and conscientiousness.
We must thank them with great Respect and deep humility and in a way both they and we understand.

Our Fathers - which art in the Heavens,
Holy will we keep thy name
We trust that thy kingdom will come
And that it shall be done on Earth
As it is in your kingdom in Heaven.
You have given us grain for our daily bread,
For which we thank thee.
You have forgiven us our trespasses
And we will endeavor to forgive those,
Who have trespassed against us.
Please lead us not into temptation
But try to divert us from evil.
For yours is the Kingdom -
The Power and the Glory -
Forever and ever. Thank you.
Amen.

We think that compliments and thanks to the creators should be humbly given regularly and sincerely. Maybe in a form like that above which is directed towards an actual living being from outer space.

When we speak of creation it should of course not only be of created "creatures". Just as much time, maybe even more, was spent on developing the environment. Creatures had to be kept in laboratories on the ships or in caged areas on land. The soil had to be prepared; the streams had to be flowing, the grass, plants and trees established before any creature could be let loose into the new environment.

The mountains, the valleys, the hills and dales the rivers, streams, lakes and Oceans, all created with such beauty and wisdom. In the beginning there was only flat Granite-like hardened Magna to work with. Soil would have had to be "brought in" water also would have had to be "brought in".

The first water acquired, when it was time, was from Mars, the red planet. The water would have had to be sucked up from the surface into space so it would freeze. This frozen lump would then have been set down here and melted and the water pressed out of the red mud that was also collected with the water. The great volume of this red water caused our first oceans to be red.

The procedure of dredging by dragging along the surface and drawing up the water and mud caused indentations on the surface of Mars, which appeared from Earth to be, canals.

This same procedure was applied in removing water from Saturn's small, icy moon Enceladus. The dredging procedure produced similar canal like indentations on its surface.

Soil, water and even moist soil were brought in "fictionally" by at least six Star Ships. These substances were deposited in a deep depression, which had been caused when Earth was being moved from one location to another. This depression was located at the point we now call the Grand Canyon. The various ships brought in the lumps of frozen mud and water and set them down. Heat and pressure were applied squeezing the water from the mud and flattening the lumps. As time went on a channel was established to allow the water to run off. The volume of the material brought in caused the channel to become quite deep. Billions of Tons of material were deposited from the various planets and their moons. There was also a great diversity of color in the great lumps of debris. When the lumps were flattened to dislodge water the "layered look" it produced was quite beautiful. Some of the smaller moons of the various planets in our solar system were pulled here and soil scraped from them. Some moons were completely consumed by our need for soil.

When the channel was cut for drainage much of the red dirt removed was spread over the South Eastern and south Western United States. The creators pumped mud up from the sea to form a land bridge between North and South America. They also had used mud as a cover up to their construction and for other reasons.

It became necessary to protect the shoreline from the erosion that the sea waves would cause. Many locations met the sea with rock but even more locations with soil. Certain rock was chosen and placed along the shoreline and atomic and hydrogen rays of energy and explosions may have been used to turn that rock and some of the soil to "sand". Today we use ultra sonic waves to break up kidney stones. The creators used ultra sonic waves to break up rock and produce sand. Many billions of tons of sand were produced throughout the globe and when water was released from the ice the shoreline was protected from the waves.

As you might have suspected though, there could have been side effects. The Earth could have been radioactive. The Gods then would have been forced to leave and return to their Kingdom to await a livable atmosphere.

For a while I tried to believe that trillions of big rocks rolling together in the surf had worn themselves down to little grains of sand. There are some things that even we idiots can't except even if this wearing down procedure had taken millions and millions of years. I had examined sand out at Jones Beach in New York one summer and sand looked more like what I thought the residue of an atomic explosion would look like than it did like a bunch of rocks that had gone on a diet and just wasted away.

When the Celestial Beings left, Science did not take note. But

Science did take note when they returned. Science noted big and sudden jumps in evolution from time to time. They're still trying to comprehend how evolution could do anything suddenly or make giant leaps of progress. SUPPOSE understands those jumps as the return from their Kingdom with the products of their technical improvements while at home. For instance they might have gone back to their Kingdom to wait for the environment to establish itself so they could release creatures into it. Another possibility could be a recession at home. Earth Project would have cost a great deal of money. If their monetary system suffered some sort of a setback there might not have been sufficient funds to support the project. That could have necessitated the withdrawal of personnel. The Celestial Beings would of course have brought the projects they were working on home with them. They would have continued those projects and then gone on to others. If these beings live forever and ever or if they are able to transfer all their acquired knowledge to a new body then they might have return to Earth a thousand years later releasing another giant jump in evolution to be pondered on by our scientists.

Upon the Creators return and then the release of "their" developments our scientists would have been astounded by the unbelievable jump in the evolutionary progress that they "discovered" in their fossil records. Like many other things it's just a matter of how you look at it.

Creation cannot be thought of as simply being the magical coming together of cells, as in evolution. Creation also means our environment, which includes our atmosphere and our weather. Weather is almost a living thing. Weather is like a giant living machine. Weather is awesome.

We have SUPPOSED that our first oceans and seas were of salt water that had been collected from evaporating water throughout our Solar System. Weather has given us fresh water, by drawing up evaporated salt water into vapors called clouds and raining fresh water. This system and the movement of air currents throughout the globe did not evolve haphazardly or randomly any more than creature building had on Earths surface. All of creation had to be triggered, had to be guided along with cause and effect.

The force of what we call nature is not only the rainy day that spoils our picnic but is a lot more than that. If you would please keep in mind that our globe was caused to rotate and was caused to do so at a very specific rate of speed. That rate of speed has had to have been monitored and controlled. Then to maintain its proper revolutionary speed adjustments have had to be made.

When a tire on an automobile gets out of balance a shimmy is noted in the steering wheel. To correct this imbalance a lead weight is placed on the wheel to get it back in balance. When our orb gets out of balance there is no way to place a gob of lead on the equator to straighten everything out. What then you ask can be done? Oh, I'm glad you asked. Should we slow down a tiny bit we would have to put some extra weight on certain spots of the globe to get up to speed. Sometimes Old Big Ben over there in England has a way of slowing down a tiny bit. The English clock tenders place a coin of some kind on a gear of Big Ben's works and that little extra weight gets him up to the second.

I know you are about to ask how we could do that so let me tell you. Since the creators can control the weather, that great machine, they set in motion the conditions that cause heavy snow in northern North America. When spring comes the melting snow and the heavy spring rains flood the Mississippi Valley and soak the soil. East of the Mississippi Valley that East Coast experiences unusual wet spring and summer weather and its soil is heavily saturated with water. Here then we have this unusually water soaked heavy soil giving Earth a tiny bit of a push getting it back to speed. At the same time all that water from the rivers of the East Coast are pouring into the Atlantic causing an increase in its volume and an increase in current flow. These conditions could keep the gulf stream and the whole Atlantic current up to snuff and on schedule.

Oh my. You seem to be concerned with all the people flooded out of their homes. You are concerned about the cost of rebuilding. Your concern is for the people who chose to live in flood prone areas. That is a people problem. The only important thing is that we are back to speed or that the Ocean currents are in order.

When we are gardening to grow food or even flowers we are like Gods to the insect world into which we invade. When our shovel or hoe come down with destructive force on an ants world they like us probably wonder why God has brought so much destruction upon them. But both the ant and we has to realize that the destruction and devastation were wrought for a necessary purpose. A purpose far more important than their homes, which can be rebuilt. Earths revolution is back on time, the ocean currents are flowing as scheduled and man has food to eat and flowers to admire.

Much like our fossil record, creation is fact. Creation is real. One can look around and see creation. Trees, mountains, lakes, grass, cats, dogs, kids, clouds are all facts and products of Creation. Facts are good when looking for the truth. Truth searchers have to be careful of that condition called belief. Belief deals with feelings, and accepted ideas and can be evasive.

BELIEVE

To believe or not to believe,
That is the question.

Whether 'tis nobler to the intellect
To accept such hap-hazard thinking

Or bear arms against their childish thoughts,
And by opposing end them?

Ah, there's the rub that makes SUPPOSE so hard to take.
For who would bear the ridicule of peers
When he could join them in their mistaken beliefs.

To SUPPOSE---To BELIEVE,
And in that belief feel freed.

Who knows what intellectual joy will be yours?
When you have shuffled off their antiquated theory.

For SUPPOSE is not a dagger that you see before thee,
The handle towards thy hand
Which maims and cuts into your mind.

But rather it is like a Lark at break of day arising
From sullen Earth to sing hymns at Heavens Gates.

I would like to take a moment to thank Mr. Shakespeare for his help with the above ditty. He just happened to walk through my Library as I was struggling with it. "Thanks a lot Bill; you were a great help."

Bill liked the culmination of this, of believing, better than something like it he wrote some time ago that culminated with thoughts of suicidal death.

Once I believed in Darwin's theory
But I found it dry and dreary.

Suffering much from cause and effect
Lacking substance on which to reflect.

If only it could stand alone
Not depending on fossil bone.

If explained in an intelligent form
Then t'would be logical and norm

Not explaining itself is tragic
It makes one think of it as magic

How did those cells first join together?
Forming skin bone and feather.

First two cells then four then five,
And there it was a creature alive

Swimming around with graceful motion
It suddenly had a sexual notion

It produced for itself a second sex being
And also eyes for ogling and seeing

But I can't find out what caused all this magic
That's the part I believe is most tragic

It's truth I think is truthfully dreary
It absolutely is a most magical theory

Suffering much from cause and effect
Lacking substance on which to reflect

A need for Darwin's evolution
Is a hand to guide its resolution

Without guidance and alone
It does not explain one bone.

Belief is a feeling that something is real and true. A feeling of trust. A feeling of confidence. Something accepted as true. Believe is to accept as true or as speaking or conveying truth. Believe also means having religious faith. When one searches for truth one looks to facts and tries to stay away from feelings and other people's acceptances. We have always also questioned that word "faith". In fact if one believed and had faith then Alice in Wonderland could very easily be believe in as true.

It would be nice if you could put aside your "FEELINGS" that something is true and "FEEL" less confidence in the "ACCEPTED" truths. Your mind would then be clear to SUPPOSE - PONDER - and CONTEMPLATE. That's the way to find the truth. Never mind what everyone else thinks...think for yourself.

If I could wipe the thought of the words extraterrestrial and alien from your conciseness it would be so much easier to communicate with you. Without those words to distort your thinking I could say "Let's compare evolution with SUPPOSE. Then I could ask you some questions. Doesn't it make more sense to have had actual beings join together the first two asexual cells and then build creatures, than to have had that procedure happen just by accident? Doesn't the fossil record of design change convince you that beings were involved in creature development? How can you accept the complexity of creature development as accidental?

To believe something it should be logical and reasonable to your mind rather than something you just feel is real and true. A belief should be acceptable to your intellect. I'd have to suppose that Darwin had a feeling that evolution was true and believed it. He also believed it would be accepted because he thought there was no other answer. The question of course was how did all these creatures come into being, both living and dead. Darwin knew the fossil record was fact and believed evolution fit in with it, somehow. Unfortunately he believed and had faith and so got it all wrong.

I was reading some scientific articles recently and was struck with a most hilarious thought. On the pages of a nature magazine was the very scientific name of a study in science. I had to laugh, it was right there in print. I read, "Evolutionary Biology"… Here were grown men who had studied biology in schools of Higher Learning and then dedicated their lives to a science that to me should have actually been called a study of a Non-existent Biology. In order that these scientists not be embarrassed by their science we would like to suggest that they call it Genetic Engineering Biology, of which they could be proud.

Darwin of course did know that the Fossil Record was fact. We know that the fossil record is fact. Connecting the fantasy evolution with this factual record was an abomination. Unthinking, but yet learned people have accepted this marriage of fact and childish daydreaming into what they call their understanding. Let us now examine this fossil record and see if you can find reason to divorce this incompatible couple.

FOSSIL RECORD

This shouldn't be a very long section, as I do not have much to say against the Fossil Record. You are though extremely lucky as I have arranged an interview with the most intelligent and the wisest man who ever lived. Because of my lack of education and inability to properly explain my theory he has condescended to answer one question for us.

"I humbly thank you sir... As the most intelligent and knowledgeable man who ever lived could you please tell us sir, if you believe in Evolution."

"Not only could I tell you, but also I will, even though that question has a two-part answer.... Evolution or the process by which something develops gradually into a different form without any special creation is probably the greatest fraud ever-perpetrated on mankind. Intellectually, this idea that the uncontrolled coming together of one-cell beings to form an organized living creature is about as stupid a belief as ever was devised by man.... As for the second part of your question I must admit that science, turning themselves into priest protectors of the above silly theory have drawn into it as proof, the factual fossil records of their science. Science has no right to link factual fossil records to the idiocy of cell development without any control or type of creation. The fact that so many of you underlings have been taken in by this combination of fallacy and fact makes me feel nauseous. Good by and have a nice day"

"Good by Sir, and Thank You so much."

I feel so humbly, that the most intelligent man that ever existed on Earth has in so few words expressed my opinion so clearly.

The only problem I have with our fossil record is that it may have been stretched back a bit to fit the theory that all this natural development took such a long, long time.

What I would like you to SUPPOSE ponder and contemplate is:

Do you truly believe the extraordinary design changes evident in the fossil record should be linked up with evolutions random and haphazard coming together of cells without design ever being involved?

Do you actually believe that there "is" a link between these two completely divergent subjects or that they could ever be interconnected?

Science has developed DNA and RNA and can use it to follow the development of various species now sitting in Museums as skeletons. Science may run into a sort of missing link syndrome during this type of research. They believe that all they have to do is find the fossils that are all out there and examine them with their DNA and they will have all the answers. We the readers of SUPPOSE know that many many fossil skeletons have been ground up in the toilet and garbage disposal devises on Star Ships along with God's body waste and have produce fossil fuel. It's very likely the "missing link" they have been after for so long never walked on Earth but rather lived his whole life on board a craft. When that experiment was terminated all evidence of it was ground up with toilet waste and became oil.

Are you keeping in mind that SUPPOSE is quite controversial. Are you keeping in your open mind that we respect and honor God even though we hint at his toilet necessities? The supernatural God we have all been used to, never required toilet facilities because he is spiritual and such measures are then "unnecessary". SUPPOSE now offers you a controversial living breathing God, an actual Living Being who uses the toilet.

Science has a great responsibility when putting together bones of a prehistoric creature. Now that we have DNA each bone must be checked while assembling these creatures for museums in order that the wrong head doesn't get put on the wrong body. Bone age also must be verified for each bone to make sure the hip and the jawbones are from the same geological period.

Back in Darwin's day I don't believe they had many fossils of leaves and flowers but those items have to fit in with his "Theory of Evolution". I haven't read much about how the single cells haphazardly formed various flowers. The diversity of flowers is as mind boggling as creatures. To create creatures the one cell asexual must have bumped sideways into each other to form tissue and then had to bump top to bottom to form fibers for trees and flowers.

I guess that's just the way nature is?

The subject of fossils is something you can get your hands on, you can hold a fossil and examine it and know it is real. But some things that are real don't leave any evidence of its existence like fossils do. Take God as an example of something real that leaves no physical evidence. Some humans believe and some do not. Lets try now to open-mindedly examine and see if we can come to some conclusion about the phenomenon of GOD.

GOD

Oh the subjects one gets to when one SUPPOSES. Since we've labeled the Celestial Beings, "Creators" that naturally led to God the Creator. I'd like you now, to take a deep breath, check that your mind is open and clear. Good, because we are now going to speak about God. But before we get serious we would like to share a cartoon we saw one time about a group being checked into Heaven at the Pearly Gates by none other than God himself. Who it turned out was a very large Saint Bernard. "Don't be embarrassed", God said to the new arrivals, "Everyone that comes up is equally surprised, it seems you've been spelling my name backwards down there for centuries. Now that was an open mind check, and if you said, "UGH" and made funny faces, you have to work on it a little more.

Concept is the only difference between the belief in God that many of you hold and the belief in God that we hold. We believe your idea is that, we are made in God's Image (a body with two arms two legs and a head on top) and that God is the Creator of all things. Our idea is that, we are made in God's Image and that God is the Creator of all things. Well now, that wasn't so difficult, was it? Old Abe Lincoln said once something about agreeing with people some of the time but not agreeing with them all of the time, or something like that. Well we did agree on what God is but we do have a conceptual disagreement of who God is.

Conceptual differences also apply to scientists, intellects and others who consider themselves Atheists. Since their only choice was between a spiritual supernatural entity from antiquity, shrouded heavily in superstition and magic or the belief in nothing, they chose nothing. We now offer them another concept to choose from.

We believe that God or more clearly, the Gods are naturally occurring Super-Scientific-Beings. They traveled here from their Planetary Kingdom somewhere in Outer Space, or you might say from their Kingdom in Heaven. We respectfully refer to them as Celestial Beings, the Creators or the Gods. We admire and have reverence for their deeds of Creation of all things here on Earth. We believe that all living creatures on Earth were created through true Genetic Engineering. We believe that all living creatures and all vegetation were designed and created by this series of Genetic Building, Engineering, and Cloning. The DNA Code was written first and Creation followed. This concept of God was developed with open-minded Imaginative Thinking. We believe in Creation by Design as the only possible explanation for our existence. We believe that the designers are the Celestial Beings from a Kingdom in Heaven rather than a magical spiritual entity from antiquity.

We do not adhere to the many hypotheses that some kind of Alien Creatures may have landed here and assisted Mankind in his magical evolutionary journey. Neither do we accept the dim-witted idea that these Aliens have come here in their Flying Saucers to use our magnificent evolutionary development to improve their race.

Now the above was a basic simply explanation of our God as a Celestial Being. It is of course much more complex. The following is a basic, simple explanation, which also is more complex, of another belief of who God is. God is a Super Natural Being who lives in the firmament, which he called Heaven. He created all things on Earth by the use of his Supernatural Powers. The Egyptians developed this concept of a One God after giving up the idea of many, many Gods and Idols. The Hebrews adopted the One God concept and based their Tribal Religion on it. This was done some ten thousand years ago, without any scientific facts to base it on and with only their superstitions and faith to support it. He is a loving and benevolent God.

We have been speaking of the Judo-Christian God. We understand that the believers of the Hindu, Buddhist, and all the other religions of the world fully comprehend their concept of the Who, What, Where, When and How of their Godly beliefs. We also realize that they and you are not cognizant of the concept of the SUPPOSE Gods character. Our Gods are first and foremost very highly intelligent, scientific beings. We use the word Gods because they are not just one Supernatural Being but rather are Naturally Occurring Super Beings. (If you would like to believe that your God had created them and put them in their Kingdom in Heaven that's fine with us.) They are experts in every field of science we know of and experts in many fields of science we could not even conceive of. They are Engineers who have split and divided continents and pushed them across molten magma. They are Biologists who have designed and built the Human Nervous and Optic Systems. They have designed and built Turtle Shells and Reindeer Antlers. They are expert Linguists who gave us our original languages.

They are expert Botanists who designed and grew Mighty Oaks and Redwood Trees. They created the Violet and the Rose. There are no limits to the marvels of creation conceived by the minds of our Celestial Beings. How do we know, you ask? Observation, reading, common sense, reason and most of all the wonder of IMAGINATIVE THINKING. Mr. Darwin's Theory was developed by observation, reading and mostly by imaginative thinking, but unfortunately he came to the wrong conclusions. Even though his logic was flawed his peers didn't let him know. He and now his disciples are actually expounding his erroneous conclusions as if they were fact, which by no means, could they be. As long as we are speaking of Darwin we would like to make the difference between his theory and the SUPPOSE THEORY, as they say, "Perfectly Clear". His propagators, have since 1847 been diligently working to fit the facts to his theory. As you read about the SUPPOSE THEORY you will find that we work to fit the THEORY to the FACTS.

Now to answer the guy in the back, who keeps yelling, "did you ever see them or talk to them, how do you know they even exist?" "NO, we have never seen nor talked with them" but we know they exist because we exist. We could not exist by the hit and miss mechanism ascribed to evolution by scientists nor could we exist by the magical supernaturalism of religion. "The only thing we can be positive of is that we can't be positive of anything." So let us say that we simply accept the existence of Celestial Beings.

This hypothesis that I have proposed includes the period when men did not record their own history and during which Celestial Beings walked among them. This new concept of how conditions were changes the meaning and understanding of many Bible stories. One can now look at these stories without supernaturalism and magic but with technically logical reason. When your thoughts include the actual presents of extraterrestrials the same printed words have a completely different meaning.

In Chapter 19 of Genesis Lot met two Celestial Beings, Angels, at the gate to Sodom. They entered Lot's home, washed and feasted and as they were about to lie down to rest, a mob of sorts came pounding at Lots door looking for the two men they had heard of. Lots guests must have looked different or strange to get a mob so worked up and to seek them out. The guests pulled Lot safely inside and it seems they established some sort of a Force Field that kept the mob from finding the door. In the morning the Beings got Lot and his family safely away from the cities that was about to be destroyed. Those saved were instructed not to look back at the City, as the beings didn't want them to see the Spacecrafts destroying Sodom. Lots wife did look back thereby becoming a witness to the means of destruction. The witness was dispatched by a concentrated heat ray turning cloth, flesh and bone to dust and evaporating body fluids, leaving only a pile of salt. Word of mouth established that the hand of their supernatural God destroyed Sodom and Gomorra. Verse 24 "Then the Lord rained upon Sodom and upon Gomorrah brimstone and fire from the Lord out of Heaven" (Or was it heat beams, lazar beams, and incendiary sulfur that rained down from a spacecraft in the sky?)

For some time we did not believe in flying saucers or their pilots at all, because we pondered, if they existed, they would have communicated with we Earthlings. Then there was R0SWELL. I read the first reports back in 1947 in the New York Times (it was fit to print). A short time later the Military came up with some story about a weather balloon. Eyewitnesses said they carried something away on a stretcher. A weather balloon on a stretcher? Very strange. The people living in Roswell knew that something had crashed. These people also knew that the Military loaded their trucks with debris from that incident. We have SUPPOSED that the trucks drove to Area 51 to try to reconstruct the "weather balloon?" that had crashed.

As the SUPPOSE Theory began to developed in our mind beginning around 1988 it dawned on us that if "they" had created Mankind and left us to seek our own destiny, they would not want us to know about them, and so no communication. This made the belief in Celestial Beings acceptable and believable.

The United Nations has appointed an official Ambassador to meet any extraterrestrial requesting to be taken to our leader. This appointment seems to indicate that the nations of Earth know something more than most of us do.

We do not consider the SUPPOSE God to be a moralistic, loving, benevolent being. We do believe they, the Gods, are scientists and their interest in mankind is scientific. Loving and benevolent would mean no floods, no volcanic eruptions, no pestilence, no starvation, no hate, no wars. We reject the idea of the Gods being moralistic. We feel that most of the moralistic demands of God written in our religious books were conceived and written by religious men, some with guidance from God. These men then applied GUILT as the means to enforce "their" morality. In their defense we must say that much of religions motivation in fighting for their morality was for the benefit and improvement of mankind. At the same time we must say that using morals and guilt was a means of CONTROL.

The SUPPOSE Gods are Scientists and their interest in humans is one of scientific observation. Over millions of years they have spent billions of hours developing and creating a self-propagating human being who with their help have built a variety of civilizations. Also with their help, we have a wealth of art and culture from various civilizations.

We said millions of years but that could be thousands of years, as we have no idea when they arrived on Earth from their Kingdom in Heaven. Earths scientists have put numbers on when creation or as they claim evolution began but we must keep in mind that those figures are based on Darwin's idea that evolution took a long, long time. We believe the facts concerning the development of life on Earth was probably pushed back by the learned men of Science to fit their theory. We also believe that the Fossil Record which science calls evolution is actually the record of advances in Genetic Engineering Technology.

There are many, many books that deal with a record of visits of extra-terrestrial beings to Earth at various periods of our so-called development. These records of visits, were not visits. They were accounts left by God showing his or their involvement in our development. The conclusions we've come to may not be acceptable to you but we feel the facts support them and that they are logical and that they are reasonable. We believe that if open mindedness could be applied, SUPPOSE might be the truth that all of mankind has been searching for since the beginning of time. Also keep in mind that our thoughts on this subject have developed over many years of imaginative thinking. As your mind is attacked by these thoughts you will have a hard time trying to digest them.

We believe the Creators are compassionate beings without being judgmental or moralistic. Since they know what motivates us better than we do they can understand our deeds better than we can. We then must be respectful towards them and glorify their deeds.

That being said, we must make it clear that SUPPOSE is not a new religion. SUPPOSE is a scientifically based theory of creation.

The SUPPOSE hypotheses could, with the proper open-mindedness easily by absorbed by any traditional religion. Our traditional religions are moralistic teachings based on stories (some authentic and some fabricated) that are historic facts, superstitious beliefs and truths. These facts and truths were written thousands of years ago when our understanding of life was quite different than it is today. These facts and truths were developed during the period mankind had many superstitions. Mankind at that time lived with superstition and fear of the unknown. With open mindedness SUPPOSE could help sort out true meanings.

Does God live in the firmament that he calls Heaven or does God live on a group of planets in outer space that he calls his Kingdom in Heaven? Thousands of years ago mankind believed their Supernatural God lived in the firmament of Heaven above and that is still the belief that prevails today.

Is God a spiritual being developed by superstitious beliefs of thousands of years ago? Is God a truly living Spirit? Is God an actual living being? Even though we are in the twenty-first century we still do not "know". We think, we believe, we have faith, but we do not know.

Our hypothesis is that God is one of many actual living beings. We developed this theory because it is obvious that the creation and development of plant life and living things had to be physically triggered and diligently controlled and could not have just happened. We also feel that supernatural creation is not consistent with logical thinking. Evolutionary creation is illogical and unreasonable. Supernatural creation is also neither logical nor reasonable. The complexity and diversity of creation makes its development by scientific beings both logical and reasonable. Keep in mind that you are with open mind. Remember that both evolution and supernatural creation lack cause and effect but do not lack magic.

SUPPOSE comprises of ideas in imaginative thinking. Read, ponder and contemplate. No one ever asked that you believe. After all you know as much about God as I do or as anyone else does. Scholars know a lot about God from the Bible but remember that before the printing press the Bible changed weekly. Open mind - Open mind. After all there were thirteen or fourteen hundred years of language translation and interpretation by the written hand, daily. Due to piety, superstition, and personal agenda a great deal could have been shall we say, compromised or shall we say changed. Whatever.

Remember now that no one here ever said, "There is no God". I do allow atheists to read this book, but I did not allow any of them to write in it.

When Michelangelo painted Adam and God on the ceiling in Rome he not only depicted Adam in the likeness of God but he also depicted God in the likeness of Man. That mistake and the same error by hundreds of other artists of depicting God in Man's likeness have imbedded itself in the minds of millions. God does not look like Man, wear a sheet, have a beard nor carry a sheep's crook. The image of God in which we are made is that of an actual living being with a body, two arms, two legs and a head. An actual living being, a real celestial being.

I believe that I told you this theory was controversial.

Did Lot actually see and speak to a real "angel" (celestial Being) outside of Sodom? Let's take a moment to SUPPOSE - ponder and contemplate Chapter 19 of Genesis. Chapter 19 says so much more than its words indicate.

First, in Chapter 18 three “men” appear to Abraham and he honors them greatly. Two of these men depart for Sodom. Abraham then began discussions with the third man whom he called “Lord”, about saving any righteous people in Sodom and Gomorra.

Now in Chapter 19 the two above mentioned men arriving at Sodom are referred to as “angels” when they meet with Lot. Though called Angels nothing is said about wings and there is no mention of levitation of any kind. Lot offers the angels water to wash their feet indicating that they must have walked there on dirt paths and roads. Lot then coaxed the angels to spend the night in his home, after he had treated them to a feast, thereby honoring them.

As the angels prepare to go to bed this group or mob comes pounding at Lot’s door looking for the men whom he had harbored in his home. The mob must have heard something astonishing or someone must have seen something that really got their curiosity stirred up

Someone must have observed men who had the appearance of men but yet were not men. The presents of these strange looking creatures spread quickly and a mob soon arrived. The crowd demanded that Lot bring them outside so they could see what they really looked like.

Lot, believing that these strange looking “men” were supernatural Angels was willing to do anything to protect them. One of the almost incomprehensible things he did was to offer the mob his two Virgin daughters to do with what they might if they would just leave his guests alone. That’s how strongly he believed his guests were god like.

The mob insisted upon getting a look at the men inside and threatened Lot with physical harm if he did not bring them out. The angels inside pulled Lot away from the door of his home, which he was blocking then slammed and locked the door.

It is then said that the angels smote the mob with blindness so they wearied themselves to find the door.

What do you SUPPOSE those “men” (angels) did look like? Their looks must have been really different from human men or the person or persons who saw them with Lot could not have worked up such a mob with the stories of their observations.

To Smote with blindness is quite magical. Setting up a force field is quite scientific and commonplace for so technically equipped as those "beings" were. We have proposed a force field as the mechanism of the "smiting". The Bible doesn't say anything about the appearance of the angels but the actions that motivated the mob certainly indicates that "appearance" was of extreme importance to the story. Well it's something you might want to think about.

We didn't discuss the last part of Genesis Chapter 19. That's the part when it seems to be acceptable to say "Vice is nice - But incest is best".

King Nebuchadnezzar in the Book of Daniel also considered appearance of importance. Daniel and three compatriots were under the control of said King. They changed Daniels name to Belteshazzar. The other three dudes were renamed Shadrach, Meshach and Abednego. King Nebuchadnezzar sets up an idol of Gold and decreed that at the sound of music everyone must fall down and worship the golden image. Those who do not do so are to be cast into the mist of a burning fiery furnace. Now Shadrach, Meshach and Abednego were Hebrews and worshipped only one God and absolutely no idols. They sort of told the king off and said if he cast them into the fiery furnace their God would deliver them out of his hands. Kings in those days didn't like being told off so he commanded the most mighty men in his army to bind the three dudes and cast them into the very hot burning furnace.

The furnace was so hot that the flames of the fire slew the men doing the casting. All the kings' men standing around looked into the furnace and saw four men inside walking around. The king ran over to the furnace and looked in and said "Lo, I see four men loose, walking in the midst of the fire and they have no hurt and the form of the fourth is like the Son of God." (The King made note of the "form" or appearance of the fourth guy.) The three dudes were ordered to come-hither and they came out of the furnace and were examined by all the kings' men. (No more is said of the fourth guy) It was noted that neither the dudes' hair was singed nor their cloths burned nor did they smell of the fire they were in. The king blessed their God for sending down an angel (the fourth guy) to deliver them from him. He also decreed that anyone who speaks anything amiss about that God shall be cut to pieces and their houses be made dung piles because there is no other god that can deliver after this sort.

We all know that if one goes into a very hot fiery furnace one gets burned and if you stay too long you die. Should you believe in magic you could also believe the above story.

Lets talk about the fourth guy. Who was he? Back at Lot's house we called an angel an extra terrestrial. That is what the fourth guy was. He was a Celestial Being specialist. His specialty was Holography. He produced a Hologram of the fiery furnace by a reflected laser beam. Shadrach, Meshach and Abednego were cast into a Hologram already occupied by an extraterrestrial who was controlling it. The real furnace was next to it. The Hologram was so real that the mightiest men of the Kings Army after casting the three into it, stepped back and fell into the real furnace. No more was said about the fourth guy because the Holography expert faded the image out and was beamed up and out of the picture. Besides having a form like the son of god the king also refers to him as an angel. What do you suppose his appearance was like?

What do you think? Was the story real? Was it magic? Was it a Hologram" What answer do you get by SUPPOSING?

Our concept of God came about by SUPPOSING. But what of the "spirit" of God. We have SUPPOSED that God is an actual living being rather than a spiritual being. Mankind we know is an actual living being, but mankind is also an actual "spiritual" being. God has given man a spirit within that is called SOUL. Biological man plus spiritual man and biological woman plus spiritual woman are what makes human beings. By SUPPOSING pondering and contemplating lets find out all about the SOUL.

THE SOUL

SUPPOSING led us to a physical being God. But what of God's "spirit". That we believe is what we call Soul. Genes control eye and hair color, bone structure, size and other physical features. There are though non-physical elements of humankind like personality, integrity, talent and other characteristics that are not controlled by the physical gene system. Those elements we have SUPPOSED are the features of the soul. We would like now, to speak about that Soul.

Science and Religion got together many centuries ago and gave the Soul to the Church. Since it was spiritual and holy, Science wanted no part of it. The church then took the Soul under their wings and developed it into today's beliefs. Generally speaking, those (Christian) beliefs hold that the Soul is with you for your life. When your time comes to "pass on" the soul is transported to Heaven (or for the wicked, to Hell). In Heaven all are reunited with, relatives, loved ones and friends, who had passed earlier. It is a lovely scenario and makes the inevitable "death", more acceptable to many.

We do not wish to knock this story. Let me tell you of a personal experience with my Ma. One day in her senior years she said to me. "I really can't remember what my Grandmother looked like, that will be very embarrassing when I meet her, again." I knew she was sincerely worried about this, so I told her that her mother, father and her brothers, John and Joseph would most likely greet her first and they would introduce her to her Grandmother. She seemed content with that. So for all those who find great comfort in the belief, that you will be with all your loved ones when you get to Heaven, please take into mind that our following hypotheses is a purely imaginative thinking process, and should be taken with a grain or two, of Salt. Also keep in mind that you are only now SUPPOSING, contemplating and pondering.

There are many names for the Soul. One is Spiritual Essence, and another is the Hindu expression of Cosmic Essence. Some think of your Guarding Angel as soul or it could be called your Spirit, the child within or even your conscience. The awesome function of this "programmable Essence" leads us to consider it to be a Holy Spirit. We will be mostly using the name Soul because it is familiar to so many people. Now we actually have no idea of what it is or its mechanics. But we are going to explain it to you now. We have the same problem with its mechanics as Darwin had with the passing on of characteristics from one generation to the next. It was years after the "Origin of the Species" that the Botanist Mendel, in 1866, discovered the Laws of Inheritance from experiments with garden peas.

These hypotheses regarding soul did not come to me all at once as reading it will to you. It was for me a slow process over many years of wondering, pondering and contemplating as it came to me piece by piece that eventually fitted together very clearly. This is why I have asked for your open mindedness this is why I ask you to slowly SUPPOSE. This is why I ask you to be a SUP- POSER rather than a JUDGE-MENTOR.

We have no idea of what kind of a process is required or how it works, that enable a Spiritual Essence to enter a biological entity. Neither do we know when this phenomenon occurs but we continue.

A newly formed Soul contains the basic tools for life. Much like a computer these pre-programmed spirits have the knowledge of how to breathe, how to suckle, swallow, cough, grip, how to laugh when happy and to cry when not happy. These newly formed spirits were somehow duplicated into existence by the Creators, as were the first newly formed biological bodies of man, and his partner, woman duplicated by cloning. Its first incarnation is very difficult because it must start from scratch to develop its feeling of self. (I think, therefore I am.) Then after several reincarnations Souls develop, with the experiences gained in each incarnation into a new Biological Body. It thereby builds its basic character, talent, ethics, personality, ego and its feeling of self, of being. It develops what science has been searching to find; it develops its consciousness. Now all that was good but we cannot forget about, greed, dishonesty, cruelty, shiftlessness, brutality, callousness and hate among others. An Essence does not make Moral or any judgments. It does not harbor any guilt. It does not reprimand or punish.

If its biological body truly enjoys drinking and/or smoking, those pleasures will be entered as good elements and carried over when the soul enters a new biological body. Our scientists will then call one a disease and the other a habit and attempt to cure both. Love and sex preference is also pre-programmed into this Holy Spirit.

The Soul is like a computer. If you put in the right information, you'll get out the right action. But "Put Junk in and Junk will come out". It is a spirit that resides in a Biological Body. A MRI cannot detect it. Lobotomies or electric brain waves cannot reach or re-program or in any way have any effect on this spiritual computer called SOUL

We are all familiar with the Biological Body, the residence of the Cosmic Essence. The Creators developed that body in Laboratories. A group or team of Celestial Science Beings created the muscles bones and flesh of the arms. Another team worked on the development of leg features. Meanwhile dozens of other teams, working separately at various locations studied the DNA and RNA plans on their Computers (which we believe may very well be within their brain) and built from those plans Nervous-Optic, Blood, Digestive, Skin, Bone, Teeth, Sex and Brain Control Systems. When each Team had perfected its project they where assembled as a body. The Body was cloned repeatedly, and corrections made where necessary. The development of these Clones where followed closely to detect any deviation from the intended results.

This section is about the Soul, but we have here mentioned the Biological Body, because together they make up the two separate and individual beings that are you, the Homo Sapien.

The Soul is from the Kingdom in Heaven. It was transported here by the Celestial Beings in an effort to create a sustained memory and have a means to instill artistic talent in the creatures that they were creating. It most likely enters our bodies with some kind of Cosmic Ray. It lives within our bodies, guarding, guiding, advising and learning the new things we program into it, by our actions. When the time comes for our Biological Body to leave this Earth and die, the Soul goes to the light and is transported into the Cosmos. It takes with it its old and its new teachings. The biological body and its brain dies and all memories of Earthly names, locations and acquaintances die with it. The spirit does not retain any

Earthly memories. There may be some sort of a time element and then the Soul enters another Biological Body. It then using all of its pre programmed and passed experiences, guards, guides, satisfies and advises its new being. It guards by that feeling we sometimes get that maybe we shouldn't go down that dark street. It guides by consulting its ethical file, so you know that an idea was John's, and refuse to take credit for it. It satisfies by letting you experience sexual passion, without having been told how to do it. It advises by the inner voice that says, "I can do it", as you ponder a promotion or new job. Almost everyone is born with an experienced Soul. It is what science searches for, our consciousness.

Every parent has great expectations for their newborn. They love it, they nurture it, and they pamper it. As it matures, they teach their pride and joy of God and of goodness. They read and talk to their love about being good and being honest. Then their "Idol" robs a store, mugs an old man, steals a car and is arrested and sent to prison. "Oh, they lament, what did we do wrong?"

Now on the other hand, lets take a successful Brain Surgeon, and his Attorney wife. Neither of them have much time to spend with their offspring. Several different nurses care for the baby. Nannies are employed until Boarding School then it was College and the University. The parents sent money and the family got together sometimes during the Holidays. The kid develops into one of our finest Nuclear Fission Scientist, a man of honesty and integrity. These stories do not seem to compute. Here's the rub that makes for life's calamity. We do not have a choice of what kind of a Soul is assigned to each new being.

SUPPOSE, there was this lovely church going, hard working family on a farm somewhere in Wyoming and they were blessed with a beautiful, healthy, boy. They brought him up in every correct manner. They loved him dearly and were rightly, proud of his Sunday school promotion and awards. Then in his early teens he began to collect World War II, Nazi Memorabilia. He met another young collector in the next town. They were extremely proud of their Nazi interest. In time they both joined a Neo-Nazi Hate Group that had purchased land and set up a Nazi community. The boy's parents were devastated. His father had fought in the war and was with the American troops who liberated a Nazi Death Camp. He had talked to his son about the despicable conditions they found there. He had spoken with true disgust about the actions of these Nazis and his conclusion that they were sub human creatures. How, We ask you, do you SUPPOSE a situation like this could occur? Please, take a minute to contemplate it.

SUPPOSE that this soul had occupied the body of a Nazi, in Germany, during the war. It would have been very proud of its uniform, boots, and armband. It would have bathed in its feeling of Superiority. It would have been arrogant about its hate, for all those, who it "knew" to be inferior. In that context it would have been praised for its contemptuousness. In that place, at that time, under those conditions, these actions and feelings would have been considered, to be good. The spirit would be very proud of its Biological Bodies actions and feelings and program them into its spiritual computer to continue its pleasure. Then the war was over. The Nazi became a private citizen. Within his brain, so also within his soul, he still strongly harbored, and secretly held his undying love for the Fuehrer, his party, and its worldly aims. Then, when that Hitler loving Nazi, died, his soul was assigned to a brand new, Wyoming citizen, and it began a new life as an, "All-American Boy."

A boy, who would be somehow urged along towards "leanings" which were not acceptable. This urge, this compulsion from some where within him, felt to him like the correct thing to do. Now contemplate that for a few minutes.

You don't SUPPOSE that could be the reason for the generation after generation of hate by the Irish towards the English. What about the Civil War organization of the Ku Klux Klan. It's now the twenty first century and there are still guys running around with sheets on and burning crosses. We tend to conclude that the above examples, of the reenactment of despicable behavior, are not always actions taught anew, as some believe, but rather are sometimes compulsions from within.

This is what we have SUPPOSED about Alcoholism. It is not a 1. - Disease or a 2. - Dastardly evil habit. Alcoholism is a compulsion towards a pleasure you had developed in passed lives. If you are an alcoholic you had been a drunk in possibly several passed lives. It is not inherited biologically. AA does not help control a habit. AA helps control a pleasure orientated spiritual compulsion that your Soul has developed while occupying its previous biological bodies. Biologically you are not responsible. The more alcoholic experiences your soul has encountered the more difficult it is to overcome the compulsion. AA does succeed and has been successful.

We have SUPPOSED that smoking is also a pleasure your soul knows you have enjoyed before and it is programmed to repeat, for your continued pleasure, this passed "good" experience. The programming of these pleasures have been done by your biological brain into your spiritual souls computer. As you know computer programs are not written in stone. They can be changed.

Do you SUPPOSE that any of the above could be true? We did ask for your open mind, and we now hope you will "ponder" on our thoughts.

Lets move on now to a more interesting subject - SEX. We have tried not to be judgmental as we have SUPPOSED, pondered and contemplated. Guided by the hypotheses that the soul was pre programmed and that it controlled sex preference our search for truth and our open mind developed the following step by step series of "possibilities" of how it might work.

The first two things that are programmed or Genetically Engineered into every Cosmic Essence, the Soul, are first their life giving necessities of breathing, suckling, swallowing, coughing, gripping, laughing and crying. The next life giving programmed addition is their sex preference. As far as the soul is concerned there are only two sexes. The SOUL never deviates from these two.

1. The Male Sex.
2. The Female Sex.

There are no other combinations as far as the SOUL is concerned.

The Male Essence is given, biologically, an aggressive sex drive. He is then pre-programmed to be sexually attracted to the Female Vagina, Breasts, Shapely Hips, Ass, and the overall femininity of its Sex Object.

The Female Essence is biologically given a more passive demeanor and a rather passionate sex drive. She is then pre-programmed to be sexually attracted to the Male Penis, Testicles, Chest, Ass and the overall Masculine Body of its Sex Object.

That's all there is to it. Sex is simple.

Why, what's the matter friend? There's a snicker on your face. There's a giggle in your throat. Wait, you're laughing. Tell me, what is your problem. You're pointing to the word "abomination" in your Bible. Speak now; tell me what is wrong.

"The deviates, what about the deviates?" Well, we did intended to speak of them but we can see that you have lost control of your open-mindedness. This requires a little deviation on our part for a moment to help you contemplate and ponder.

Ten or so thousands of years ago the mysterious same sex phenomenon was conceived by the Hebrew Scholars to be a "choice". They could not tolerate this choice, because they were a tribe and without copulation, the tribe would not increase. So they preached Abomination. They wrote Abomination in their books and said, God said so. Two thousand years ago the Christians look in the books and agreed. It's an awful choice. Without copulation there would not be a next generation to put "contributions" into their coffers.

As far a Sex being simple - it is. One Male Sex and One Female Sex.

This hypothesis we are getting into here is new to you. Years ago I watched TV shows on which those called deviates claimed they were either men or women who were trapped in the wrong bodies. One show I remember was a guy who said that when he went to kindergarten and the teacher instructed the class to line up with the girls on one side and the boys on the other. Even though he was dressed as a boy and had been told he was a boy he lined up with the girls he said, "because he knew he was a girl."

I have had some time to wonder ponder and contemplate these thoughts. Watching an interview of both a male and female recipients of sex change operations I was able to conceive how strongly they felt about their predicament. I was able to apply compassion, empathy and understanding to their quandary. Go slow and try to understand without making judgments. Try to SUPPOSE.

But just for the sake of argument let us now SUPPOSE, that a male body is assigned a female Soul. What would happen? We'll tell you what would happen. The male's biological body and brain would be controlled by the sex preference of the "female soul" and be sexually attracted to... a male body. Since sexual preference is controlled by the Spirit, then this Spirit would, naturally, compel its Biological Body to be attracted to "its" sex object...the male body.

Try now to contemplate that for a moment.

(We understand that you may not at this time believe a word of these written words. Hogwash, Utter Stupidity, Ungodliness. That is the reason we ask for open mindedness and contemplation. We ask for your logic and your reason. We ask that you SUPPOSE.)

It would then follows that if a female biological body were assigned the Cosmic Essence of a male spirit that biological female would be sexually attracted to a biological female. It's that simple. By the way, Choice is the kind of ice cream you wish in your cone.

A male experiencing this phenomenon for the first time would "know" he was female and would enjoy dressing accordingly. He might, nowadays, undergo hormone or sex change surgery to make him what he knows he is. Psychologists would name him a Transsexual and attempt to cure the problem.

This experience could be encountered for several incarnations in a row. The strong compulsion to be of the opposite sex might diminish some and he could be quite content simply dressing as the opposite sex, on occasion. The Psychologists would call it a psychological abnormality and label him a Transvestite. At $175.00 an hour a good Doctor would work diligently to cure the problem.

When this is a "first happening" for a Female, she will dress as a man, consider, Hormone Therapy, a sex change operation and maybe be labeled a "Butch Dyke".

A biological body, containing a Cosmic Essence that has experienced both Hetero and Homosexual experiences would be labeled as Bi-sexual. That person would not be an abomination because they could do both. As long as they where able to perform in a

"Normal" fashion there would be no reason to cure them. Society would respect them as a higher class of abnormality.

By the way, if pre planned sex preference should ever be proven, then we would have to concede that no church and no well meaning Preacher have ever "cured" any Human Being of its God Given Sexual Preference.

Well, Hunky Dory, the above was a problem that has plagued Philosophers since the time of Plato. We thought Freud had the answer but that petered out. Then we had Psychologist, Psychiatrists, Psychoanalysts, psychotherapist, Behavioral Scientists all looking for the answer. Religious Priests, Rabbis and Bigots all added their opinions over the centuries. It may also have caused you some consternation. We seem to have answered this problem of, Sexual Abomination and Sexual Deviation, quite nicely and all we did was to SUPPOSE.

Like myself many of you are asking yourselves, why? Why would God put the wrong soul in a biological body? One possibility that we might consider would be that God, knowing how strongly he had developed our sex drive, also knew that there must be a way to satisfy it. A female soul experiencing a life in a Male body would learn (or be programmed) to be more sexually aggressive. The more aggressive female soul when returned to a female being might develop into a whore or prostitute. This could be Gods solution to a healthy sex drive and a weapon against rape.

There is also the question of whether the sex role swap is a haphazard selection or might there be a gene within a family that triggers it. We do not know. We do know of a family that had two brothers who were Gay, one family with a Gay brother and sister and of a Gay guy with a Gay Aunt. These examples are not enough to prove a point. We also feel that there are very many families that have only one Gay in their History. This area could, with some difficulty be researched. The difficulty would be the Gay "secret" and cover up that would conceal the facts.

The creators are all scientists and one of their sciences is psychology. They could have done the soul switch, just to find out how we Humans would react to the phenomenon. We might venture to say that if it were a test we would have failed. The Creators may have also wondered how their pre-programmed passive Female would react when controlled by the aggressive Male testosterone. Also how the Male pre-programmed Aggressiveness would handle the passive influence of the Female Estrogen. The Celestial Beings are taking notes, preparing papers, writing journals and discussing our progress. Should the time come that all of their experiments are completed and they reveal themselves to us we can all read their conclusions. So try to spend less time on the criticizing and more time trying to understand that which we consider a problem.

Speaking of problems here is one that has plagued us for some time and you might want to ponder it. Why does everybody have nipples? Our first idea was that a biological body was conceived and then a soul was assigned to it and then its sex organs developed. The last part that is created is the small triangular groan area that contains the proper sex organs. The last to develop sex organs could explain why everyone has nipples, because one wouldn't know till the last minute if they were needed. But it seems more reasonable that the mass cloning that would have been necessary was accomplished by cloning all bodies with nipples and without sex organs, which when added completed the clone.

You might remember that in our earlier Science-fiction story we suggested that our Celestial Beings might be asexual but were able to reproduce themselves. SUPPOSE that when creating man in their image they had wanted man to also be asexual and also like them, pass on all acquired memory. Clones would have been prepared before the reproduction procedure was developed. But unfortunately their technical genius could not duplicate the complexity of their reproductive system in man. Some of the nippled clones were made to be masculine and some to be made feminine

We know that science is looking for a physical reason for the malfunctioning of sexual preference and some are sure that further study of the Double XX Chromosome of the female and the XY Chromosome of the male will lead to the answer. But listen scientists - Sex Preference is controlled by the pre-programmed spiritual essence called soul. "Sex preference" has nothing to do with biological bodies, genes or chromosomes; I know that all scientists are now happy to know that.

Since science never gave any exploratory time to the soul because it belonged to the church they credited everything to biology. Science has assigned genes to every aspect of our being. Lets Suppose that Genes only affect biological things, like hair and eye color, bone structure, nose shape, things like that. We would have to then let the soul take credit for suckling, sex preference, talent, character, personality and things like that what do you SUPPOSE? One gets the physical and the other gets the physiological

I wonder how I got to say so much about a subject of which I know so little. My sole experience with the soul was when I went backpacking. I sort of invented a guardian angel type of soul that I called "Joshua". Being alone and walking in the woods for 15 (and one time 21) days at a time one needs a little imagination to help pass the time. I managed to imagine (visualize?) Joshua up ahead in a tree or on a rock and I talked to him urging him not to let me get lost but to keep me on the trail. We spoke about the weather, what I might eat for supper or breakfast or about picking a campsite for the night. Oops, I think I may have just made a big mistake. Those of you, who thought I was nuts, are now sure of it. Those who hadn't thought I might be nuts, have now.

Since writing about the soul I wondered how long it lived or if it lived forever and ever. The wondering brought me back to Genesis Chapter 5. It's called "the book of the generations of Adam" and it got me to supposing. SUPPOSE I pondered that the accomplishment that God (a Celestial Being) was most proud of was the development of the Soul and its incorporation into the cloned biological bodies he had created. You may have taken note that in Verse 7 of Genesis Chapter 2 the Lord God "formed" man and in verse 22 it states that from Adams rib the Lord God "made" he a woman. With that information in hand we all see quite clearly that both Adam and Eve were "parentless". I say that since cloned humans are individually created without having had parents. Thusly we must conclude that the story of Adam and Eve is to celebrate the first time that Humans were allowed to mate and create life for themselves.

Genesis Chapter 5 does though call itself the "generations" of "Adam". SUPPOSE the name (as I named mine Joshua) God gave to Adam's Soul was also "Adam". SUPPOSE that verse 3 of Chapter 5 when stating that Adam lived an hundred and thirty years and begat a son in his own likeness.... And called his name Seth was actually speaking of Adam "soul". Verse 4 states the days of Adam after he had begotten Seth were eight hundred years and he begot sons and daughters. Verse 5 concludes that all the days that Adam lived were nine hundred and thirty years: and he ("Adam the soul") died.

Prior to SUPPOSE I had very little "faith" that the biological human called Adam lived nine hundred and thirty years and even more to the point. I just plain didn't believe it. Now on pondering and contemplating the thought of the "Soul Adam" (a living spirit) living nine hundred and thirty years in many different biological bodies (all also called Adam) which were always young and able to "beget" the magical-ness of the story subsides.

In Verse 6 we find that Adam's son Seth('s Soul) lived an hundred and five years (in a variety of different biological bodies all virile and all adult clones) and begat Enos. After Enos was born Seth (the Soul) lived eight hundred and seven years and begat sons and daughters (with these constantly renewed physical bodies). Seth's Soul died when it was nine hundred and twelve years old

Enos begat Cainan at ninety years and his total years were nine hundred and five.

Cainan begat Mahalaleel at seventy years and his total was nine hundred and ten.

Mahalaleel begat Jared at sixty-and five years and his total was eight hundred ninety and five.

Jared begat Enoch at an hundred and sixty and two years and his total was nine hundred sixty and two.

Enoch begat Methuselah at sixty and five and his total was three hundred sixty and five. Methuselah begat Lamech at a hundred eighty and seven years and his total was nine hundred sixty and nine (Methuselah's Soul won he beat out Jared by seven years................Three cheers for Methuselah)

Lamech begat Noah at an hundred eighty and two years and his total was seven hundred seventy and seven.

Noah begat Shem, Ham and Japheth when he was five hundred

years old. (His soul was five hundred but his newly cloned body was young, strong and sexually active.) Noah his family and all the animals entered the ark when Noah(‘s Soul) was in its six hundredth year, his total was nine hundred and fifty.

I really tried to keep my thoughts separate from the Bible Truths, as you know them. Genesis doesn’t say anything about how long a soul lives. Insurance companies try to figure out how long people will live but they have birth certificates and death certificates and many other records to aid them. All I have is my Wonderful Wondering...Stimulating Supposing...Puzzled Pondering...and Confused Contemplating, all of which I use diligently.

There is a slight possibility that I might be wrong in some small conclusions I’ve come to but it is highly unlikely as you, I’m sure, have noted in your reading up to this point. After all I’ve never been wrong before.

By adding up the total years of the ten above mentioned from Adam to Noah and dividing by ten I have deduced that at that time a soul lived an average of eight hundred fifty seven years and six months. Isn’t that something you always craved to know?

Yes... Yes... I know that could be a stupid ignorant foolish conclusion. All I can say is that to believe those men actually and biologically lived all those years is even more heavily weighted with ignorantly foolish stupidity and is further laced with considerable magic and superstition.

First I wondered about some words in Genesis and then I SUPPOSED. God the Celestial Being I feel must have been quite proud of the creation of human kind. But I pondered that the placing of a spiritual essence within humans which could work with that human’s biological brain was an even greater miracle of their creation genius. I concluded that much research over long periods would have been required to keep track of the experiment. Adam who knew Eve would have in time died. The Celestial Beings would have then taken a new clone, called it Adam and placed the soul of Adam who knew Eve into the new Adam and called that soul Adam. Then another death, another clone, another soul transfer and another Adam both human and spiritual. Eight or nine hundred years later the spirit essence dies. Hundreds of years of close observation of an experiment of creation. Each of the clones was a full-grown human. Waiting for babies to grow through childhood was never necessary.

No one ever said to believe me. We are supposing, pondering and contemplating. We are with open mind and have just received some food for thought, something for our minds to chew on, don't just spit it out, digest it.

Before Adam and before Noah there were other beings that were monitored and whose lives were controlled by their creators. They were taught new things and taught how to draw pictures and then the biological bodies were terminated and the soul transferred to a new clone as the souls abilities were perfected. Those pre-Adam and pre-Noah beings are known to us today as cavemen.

HOMO SAPIENS

I didn't know that Homo Sapien was the name of my species when I was in the sixth grade but at that time they taught me about the first beings, (Java Man, Peking Man, Heidelberg Man, Neanderthal Man, Cro-Magnon Man) the cave men. Cave men lived in caves because they didn't know how to build houses. Sometimes after an electric storm with lots of lightening they left their caves and went into the woods looking for fires started by that lightning. When they found fire they pushed it onto flat rocks and carried it back to their caves, adding wood fuel. They had the woman keep the fire burning and went again to the woods with sharpened sticks and rocks to kill some animal and bring it back to the fire. Cavemen had accidentally learned by carelessly dropping a piece of meat near a fire that the meat tasted better when "cooked". Whenever they found fire in the woods they tried to keep it burning in their caves while out hunting for meat so they would not have to eat the meat raw. Cavemen also had to keep fire burning as the weather became colder, as there where few lightning storms while snow was on the ground, and even fewer fires started by lightning.

Cave women soon taught themselves to make thread and sew together skins to help keep warm when fire was not available during the cold season. By accident cavemen discovered that striking steel on flint caused a spark, which could in turn start fire on tinder. Cave men studied the wild animals in the forest and taking ash or charcoal from their fire drew pictures of the animals they hunted. They had no language so pointing and grunting was necessary to make their wishes clear.

I learned all that in the sixth grade. I never had reason to disbelieve it. But what a bunch of hogwash it was.

Now that I have been able to step back, SUPPOSE, ponder and contemplate I have come to a quite different scenario.

God would not want his newly developed experiment to catch pneumonia so I feel certain he would have assigned each one to their own room with toilet facilities and air conditioning inside a Star Ship.

God linguists would have taught them their language as soon as their voice boxes had been successfully genetically engineered and installed. When not occupied in learning classes they would have watched educational programs on their communicator screens.

Learning Classes of four or five cavemen would be conducted by, say a celestial being geologist specialist. Taking the group into the wood and explaining to them how to locate and identify the proper rock to use when striking the proper flint. As each cave man chose his material the teacher would have instructed them on the proper procedure to produce fire. Learning how to find water in the forest would end the morning segment and the teacher would pass out the seeds, nuts, fruits and vitamin pills on the menu for lunch.

After lunch the geologist would take them deeper into the forest to find rock for weapons and tools. They would learn arrowhead making and spear point development. The cavemen, Russell, Charlie, Oscar, Pete and George would learn bow making, arrow making and become proficient at producing axes. They had to find their way back to camp at the completion of their field trip. On the way, using their new weapons they hunted meat for their supper. The teacher would tell them about the position of the sun at different times of the day and how to tell the direction north by the moss on the tree trunks. By noting the direction north and observing the route of the sun during the day they were better able to find their way.

Upon arrival at their cave Russell called his mate Agnes (who had spent her day in dressmaking class) and proudly showed her his flint and steel and explained the method she could use to start a fire should hers go out. Carl his friend, standing near by, had been at tanning school this week and was looking forward to the stone tool and fire class he was scheduled for next week.

Charlie, Oscar, Pete, George with their mates joined Russell and Agnes around the fire all talking loudly, at the same time and all demonstrating their proficiency at starting fire with flint and steel. The fire was large and hot and cooked the meat quickly while the group in festive humor feasted on the fruits, nuts and seeds some of the girls and their nutrition teacher had gathered during the day.

At Ten PM the "no more noise" bell sounded and as they had been taught they said goodnight to their friends and retired to their rooms for a good night's rest. The morning would bring a nice hot shower clean skins to wear and a hearty breakfast. Then another day of glorious learning taught by their creators.

Oh the life of a cave man was wonderful.

Sometimes on Saturdays (a rest day) when they got their weekly physicals and had their vitals taken or on rainy days they had art classes in their cave. Russ loved art class.

It is within my understanding that things did not go just as I represented them in the above story. I do get so carried away. On the other hand, you, now being so thoroughly evaluated on the true origin of the species were cognizant before you read the above story that the Creator God would have walked and talked with his creations. God also taught them what they would need to know in order to survive.

Earths Scientists also being Homo Sapiens generously gave the cave men credit for their ingenuity in discovering and inventing things on their own. Being of the same species the cave mans cleverness helped inflate the scientist's ego, which is always of great importance. As time went on we are taught, Prehistoric Man became extinct. As time went on we now know that the experiments with Pre-historic Man were terminated when the Homo Sapien was developed.

I think we all know that the story of Noah is a bit out of whack. All you need is just a tiny bit of logical thinking to realize that even though Noah wanted to do as God commanded there was no way he could have gotten together two of each species. Maybe there was a lot of rain and a flood and maybe he built a raft for his family but the two of each species is beyond what reason will allow. We are in the twenty-first-century and we are still finding new species. Noah couldn't fit just the species discovered in the twentieth century into his ark. The termites and the boring insects would have sunk his ship. It just makes me wonder how stupid the men who wrote it thought we were. I guess in the years 500, 1000 and 1300 people just believed whatever was read to them from the Holy Bible. Fortunately for us here in the twenty-first-century we are too sophisticated to just believe whatever is read to us.

(Would you like me to read that over again for you?)

Civilizations dawned in the river valleys of: Egypt, the Mesopotamian crescent, India and China. We will now, arbitrarily and unilaterally place Adam his descendents up to and then beyond Noah and the team of Celestial Beings working on Adam's particular Semitic Homo Sapien species into the Jordan River Valley. A length of about a hundred miles of the valley was made fertile. Thirty-mile strips on each side of the river were given the same treatment; the surrounding areas were undeveloped deserts. The available flora was planted throughout the area. Prefab houses and laboratories were moved in. A team of twenty- six "Beings", fourteen inactive clones and another twenty seven that had been activated and equipped with souls were immigrated into their "New World". The refinements necessary in creating women from man's protoplasm were taking place in the Star Ship Laboratories. Eve was still being developed. As was also all the beauty love, understanding and tenderness we know of even to this day to be 'WOMEN'. Oh My - I'm embarrassed - Oh Gosh - Gee - Thank you Ladies your welcome ladies you are most humbly welcome.

Lambs for wool and string, goats for milk and cheese, swine for meat and for lamp fat (fuel oil), oxen for food and plowing fields, asses for transportation and hauling, chickens and roosters for eggs and meat were all immigrants into this complete world of the Jordan River Valley. All of them having been elsewhere created, developed and domesticated by genetic engineering. (Sort of like a second influx of "aliens" on earth).

This was during the period that man had not yet been taught to write and maintain a history, so God readily appeared to and with him. God worked in the fields with man developing and teaching farming methods, God taught the principals of and helped man with the construction of buildings. When the female species was developed and female clones activated God taught them how to spin wool for yarn and helped them construct looms for creating cloth. Farming, weaving, building and food preparation were by these experiences, programmed into the souls computer. When death came to the souls protoplasm home and it was assigned a new Homo Sapien. It was then able to by its past experiences and in its own subtle way make the tasks encountered by its host seem a little easier

And it came to pass, as you know that clone Adam and clone Eve were permitted to copulate. Since there were no temples or churches at that time the copulation was done out of wedlock. Everything seemed to go OK so other clones where allowed to mate. Over the (thousands?) years the children of the clones mated with other clone-parented children and everything was going along without a hitch. The population of the Jordan Valley community by the time Noah was born was not excessive, as God had the inhabitants sex drive restricted as a method of birth control. Many of the babies that were born were taken to the space crafts laboratories for dissection and study. This procedure was a scientific necessity.

Please remember you are reading imaginative thinking hypotheses and that you are doing so with an Open Mind while searching for truth. A little deep breathing now and on we go.

One day God told Noah they had to talk. So sitting beside Noah's herb garden next to his house they had a serious conversation. The medical lab had informed God that test had indicated that almost the entire population of the valley had developed a hereditary and contagious cancer cell defect. Neither Noah nor his family had contacted it. Twelve others were also safe. The Celestial Being had drawn up plans which he submitted to Noah explaining that he was to build an Arc with the help of his family and the other uninfected twelve men and woman. Those that were to enter the Arc were given vaccination for protection.

The Arc people having been told of a great flood worked hard hewing trees, sawing logs, constructing, pouring pitch, and gathering food for themselves and for the animals. Noah had been instructed to gather together two of each animal and take them on the Arc. Noah followed the Lords instructions to the letter. Noah rounded up one male and one female of every animal in his world, the Jordan River Valley. Noah herded to the Arc, lamb, goat, swine, oxen, ass and chicken. Six different animals he herded them there.

Construction was completed and the rains came. The Jordan

River was dammed by the celestial beings; a force field "wall" of slimy material was put around the valley community. The area within the walls filled with water drowning thousands. Or maybe just hundreds.

You may have been under the impression that god wanted all the worlds animals in the Arc, which of course would have been impossible. God wanted all the animals in Noah's world saved as Noah would need them on the Arc and also would have them when the water receded. One hundred miles long and sixty miles wide, that was Noah's world he knew not of any other lands.

Sometimes silly stories have serious meanings. Part of our silly story was to suggest that accepted theories on cavemen and their development might require some rethinking. Mankind has found a few fossils of cavemen, some artifacts of theirs, cave fire pits and animal bones in their caves. From these they have used their imaginative thinking and developed their own silly story of how things where back in the old days. You and I now know that mankind's concept in regards to caveman's conception in the first place, is seriously flawed. With this flawed concept of Homo Sapiens conception one can see how the facts we had learned "might" actually be silly stories. If the Creators had only taught these early Homo Sapiens to take notes, all this confusion wouldn't now be troubling us.

Another part of our silly story was to suggest that maybe Noah might have been confined to an area smaller than we had supposed when we read Genesis. While the world itself was being developed the Creators took small sections of various river valleys to improve and in which to develop various Homo Sapien species. The Humans living in each of these areas only knew the world as the place they lived, as the place that was their home. During this period an ever-expanding worldwide environmental development continued until the world we know today, existed. We are only SUPPOSING that at the time of the flood Noah's world might have consisted of a few hundred people and six types of animals.

Long before cavemen other developments had occurred preparing land areas for the future. Continents had to be moved and intricate measurements of crust stability and depth had to be calculated. Everything did not always go as planned and sometimes even God ran into a tragedy.

THE TRAGEDY

SUPPOSING led me in many directions. It led me to ponder geology and the why of India crashing into the land of Asia creating the Himalayas. The how and the why of the Alps were contemplated. We studied the World Atlas topography of all the mountains. Why the Urals were straight and what could have caused them since they are in the middle of the Eurasia Tectonic Plate. It was then that my imaginative thinking kicked in and I thought of a series of events that could make all those problems of geology seem real and also something we could have fun with.

From somewhere in my Religious upbringing the phrase "Thou shall be perfect, even as thy Father who is in Heaven", comes to mind. We have heaped praise on our Celestial Being as being Perfect Beings, even to the point of infallibility. We must now though admit that, gloom; error, miscalculation and tragedy can enter the lives of any living being, even the Gods. If we just tell you about the day that the Earth almost split in two it might not properly convey the horror of the tragedy. There was once a TV Program hosted by Walter Cronkite, called, "You Were There". The show dramatized some factual event in History, as though it was just happening and you were tuned in on it. So with your, indulgence and open mind we would like to present a non-factual Science Fiction episode which is based on scientific fact but is of pure Imaginable Thinking. So, 1-2-3, here we go into the SCIENCE FICTION MODE:

Commander Joseph Hovar, who had commanded the Star Ship, The Searcher, was put in charge of what the Council had named "Project Earth". Commander Hovar, was an accomplished Engineer, and had the adoration and respect of the Thousands of Celestial Being under his command. He was though an Engineer first and did not require to be called Commander. He preferred to be known as Joe. Out of awe for his genius and personal respect most of his crew used Commander Hovar, sir, when speaking to him.

On this particular day Joe had taken charge of the largest Star Ship that was working over the coast of Africa. A lazar or pulverizing type of sonic beam had previously executed a cut to separate South America from Africa. But as pressure was applied to South

America to push it West, the cut between the now two continents held. Joe knew it had not been made quite deep enough into the solidified Magna by his second in Command, Albert Lau, in the previous weeks that Lau had worked on the project. Rather than delay Project Earth, Joe calculated that by using an inordinate amount of pressure on the African coast, holding it in place, he could have another ship using similar excessive pressure push South America, West. It worked and the continents were separated.

(As a side effect of this unbelievable pressure, diamond where formed in Africa. Since we are taking this little break we would like to say that we will be using Earthling Names for various geographical locations out of a desire for clarity. The Celestial Beings would have used numbers to designate geographical areas, like Area 27, for Africa and Area 28 for South America. We feel we can maintain a higher degree of clarity by using "our" names. We think they may have called a part of North America, Area 5l, but we're not really sure.)

Joe was truly pleased that his calculations had been right on the mark. What a great day, he thought. He took his crews congratulations humbly and thanked them for their co-operation. Now that the cut was completed he turned the controls over to a technician to continue the tedious movement of the continent in ten-mile increments. His Command Vehicle was on flight deck seven, so he took the elevator down entered his vehicle and headed back to Headquarters. Unquestionably he was going to have to have a serious talk with Albert. Joe did not like to get involved in politics. He never actually approved of Albert as his Vice-Commander, but since Albert was related to the Head of the Council Joe deemed it more expedient not to express his opinion. Politics aside, Joe could not tolerate the sloppy miscalculations exhibited by his Vice-Commander Lau in the passed few weeks.

Upon his return to Headquarters he inquired about Albert's assignment for the day and was told that Vice-Commander Lau was aboard Star Ship Nine assigned to the separation of Asia from Europe. Joe instructed his staff to tell Lau to come to him, as soon as he returned to Headquarters. He then sat down at his computer to go over all the mathematical calculation dealing with placing Asia into the center of the Pacific Ocean, once it has been separated and before water was brought in to create that Ocean. Though quite complicated in reality, the mathematical calculations seemed simple and straightforward. The cut was to start at the Arctic Circle pass over the now island, called Novaya Zemlya. Then straight down the Ural Mountains, (and they say there are no straight lines in nature) along the shore of the Caspian Sea. Cutting through the Northeast section of Iran and down to about the location of Karachi. From this point it was to follow the then, shoreline in a circular curve to the area of Calcutta. From Calcutta it would pass through the center of Burma travel north of Hanoi to the area of Hong Kong. Phase Three would be the subduction along the Chinese Coast to facilitate the move. Phase One was the North to South cut from the Artic Circle through the Caspian Sea, which had already been completed. Vice-Commander Lau was now working on Phase Two between Karachi and Hong Kong.

Joe still felt good about today's success and headed down to the Rec. Room to shoot the breeze with some of his crew, and relax a bit.

Lau had been seething with anger since Joe had explained to him this morning just how he had made the mistake in his calculations. Lau was a very ambitious politician. He had wanted the job of Commander, but his relative, the Head of the Council, didn't feel he had the experience necessary to command such a large contingent of Beings. If he had gotten the Commander Position it would have insured that he would some day, soon, have the power he lived to posses, Head of the Council, the most powerful being in the Kingdom.

When Lau had volunteered to Command Phase Two for the day, he did so to prove to Joe his worth and efficiency, and to have good reports being sent to the Council. He planned on asking Joe not to reveal anything about the African incident in his reports.

There was a requirement that before any cuts were made 500-mile crust stability checks on each side of the cut had to be performed. That though would take many weeks to accomplish, and he wanted to prove himself to Joe, now. He consulted the old records made years ago by The Searcher Party that had remained here when Hovar had gone back to report to the Council. Those reports indicated that the Crust in this area was stable. He knew that since then, Earth had been moved to its present position and its daily revolution had been accomplished but he made the decisions that those procedures would have had no effect on Crust stability. Since he did not want another fiasco, like the African one, he adjusted the cutting mechanism to go just a little deeper than he had originally calculated. His crew was not happy about all this because until he had come aboard this morning their intentions were to make the 500 mile stability check, as required. Vice-Commander was in charge and the cutting procedure was initiated.

Lau was personally manipulating the cutter and everything was going as planned. He felt smug about his accomplishment as he neared the Calcutta Area. He was visualizing the look on Hovar's face when he told him about the weeks of work that he saved. He planned on saying "now that's efficiency, right Joe, when it happened.

At his Court-Martial the crew testified that he screamed and kept shouting, "Oh My God, Oh My God". A short time later with tears still flowing down his face he grabbed the emergency communicator and yelled "Hovar, Hovar we have a problem".

Joe, in his now completely relaxed mode, first, could not believe that one of his fellow beings could be exhibiting such an uncontrolled emotional outburst. He quickly snapped into his Commander Mode and demanded to know what the problem was. Lau was hysterical so Joe called for the First Mate of Star Ship Nine to take the communicator and explain to him what had happened. Clearly and concisely the First Mate explained how a piece of the Crust about half the size of Australia had broken off and sank into the Magma. The piece of Crust was slowly moving west and was ripping into the surface from below causing large cracks as it slid and tumbled along beneath the land. Hovar immediately issued a Mayday Alarm and order all Star and Work Ships to cease whatever they are working on and reconnoiter immediately at the problem area. He virtually flew to his Command Vehicle and in a short time had landed in Star Ship Nine and stood in the Control Center watching the screens as the Earth was cracked and the tumbling Granite poked though the surface at regular intervals. It slid on its full surface under Afghanistan and Iran smashed along and creating a jumble of dangerously deep and wide cracks. It twisted and its edge almost broke through in Turkey. It then completely broke through the surface at the Swiss-Italian Border leaving a giant gapping hole of fuming Magma that began bubbling out. It ripped through again at the French-Spanish Border exposing the inners of Earth. The Earth shook, the Earth trembled, the Earth Quaked. The Planet was experiencing the first Earthquake.

Before the under crust bulldozer could reach the new and thin Atlantic Bottom, Hovar ordered a Work Ship with Force Field Equipment to get above it and force it down and keep it down as it moved West. He ordered another Star Ship with the lazar like cutting beam to go to the middle of the Pacific. They were to cut a large hole in the crust and the Work Ship was to guide this offending Great Rock to that hole. He hoped then to bombard the great rock with Electro Magnetism to such an extent that it would be drawn down to the more metallic area near the center of the Planet. It arrived at the hole on schedule and during the bombardment procedure it fractured and the larger of the two pieces continued on its journey Westward. The smaller of the two sank down into the Magma. Efforts continued to force the journeying piece of the great rock down away from the bottom of the crust. Joe ordered the whimpering Albert to be medically sedated and he was placed under arrest and confined to his quarters, under guard.

Joe then turned his full attention to the three most pressing problems.

1... An unbelievably large hole in the crust South of Nepal.

2.... A smaller hole south of Switzerland.

3.... A serious rupture on the southern French Coast.

No l. was handled by having a continent size piece of land cut quickly out of Antarctica and with all the power available pushing it North towards the largest hole. As it arrived, the edge of the break was lifted slightly and the North end of the sub-continent was sub-ducted slightly so it easily slipped under the existing crust and was pushed several hundred miles North thereby putting a stopper in the hole. After the two other problems were solved, ships came back and with Electro-Magnetic Black Hole Mimicking Force reached down, grasped the sub-continent and pulled it up into the outer crust, forming a zipper like seal that stabilized the area. A lot of crust and magma was pulled up creating some pretty high mountains.

No. 2. A rectangular shaped piece of Africa was cut out between Tunisia and Benghazi, in Libya. This piece was also moved rapidly north and sub-ducted as the South end of Switzerland was raised slightly and in the same manner pushed under and north. Later, the same Electro Magnetic Force was use to zipper it up and stabilize the stopper. Not as much crust or magma was pulled up here, as the hole was smaller. Enough though to create the Mighty Alps.

No. 3. At the time of this tragedy the entire southern coast of France bordered on the Atlantic. Though this was only considered a rip or rupture but it was so close to the newly formed and very thin floor of the Atlantic, drastic steps had to be taken to stabilize it. Morocco lost more than half of its land as Spain was cut from it and pushed north to seal the rip and extend out into the Atlantic to protect the fragile Ocean Floor. No stopper was needed here but the Pyrenees is the zipper that holds things together.

Albert Lau was stripped of all his titles and as Citizen Lau was tried by the judges of the Judicial Ethics Committee. He was found guilty of putting his personal greed before the good of the Kingdom. Lau was sent to an outer planet to work as a laborer for life, deep in a mine. Upon his death his Cosmic Being was to be destroyed. The court ruled that his relative, the Head of the Council, had used poor and unethical procedures in appointing Lau as Vice-Commander and was removed from office. His entire family lived forever more in obscurity and disgrace. Had Lau been appointed as Commander of "Project Earth" the Ocean Floor of both the Atlantic and Pacific would most likely have been torn asunder; killing all beings that were working on the Planet surface. Earth would have been split in two, exploding into oblivion.

Our friend Commander Joseph Hovar, the cool headed, knowledgeable leader that saved our Planet was greatly honored and in time became one of the Greatest Council Heads of the Kingdom.

And so ends our theoretical imaginative tale. Could there be even an iota of truth in this once upon a time tale. Well for a No. 1. Science tells us that a piece of Antarctica did break off and move very rapidly north, by some sort of natural under the crust current. They also state it crashed violently into land with such force it slid under the surface for several hundred miles pushing up the Himalayas with the same kind of magical, Oh we're sorry, we mean, Magma Force Current that had pushed the Sub-continent North. For No. 2 there are a more or less a line of Mountains between India and Italy. A No. 3. Maps show that there is a missing rectangular piece of Africa in the Tunica, Libya area. Also if you visually pull Italy out from under Switzerland and square off the boot a little, twist it from its Vertical Position to more Horizontal Position, we think you will find it would pretty much fit into the missing rectangle. As a No. 4. Science has conjectured that the Eastern South American Coast fits quite nicely into the Western African Coast. With a little conjecturing on your part, slide the East Coast of Spain down to the West Coast of Morocco and see if this jigsaw puzzle piece seems to fit. Spain does protrude out into the Atlantic (as if protecting a thin Ocean Floor.) No. 5. There is the consternation of science of how such an in- ordainment amount of Lava was expelled under Hawaii. Hey guys, it was the big hole we spoke of and the necessities to pull up and pile a lot of weight on that hole to seal it up.

As a youth our reading, educated us to believe that Mountains were created by Violent upheavals and volcanic action We do not now, remember the words Science used, but we read it as Fact. Now we think we know how this fantasy came to be. Let me tell you a story.

"Once upon a time, two Geologists, Sheridan and Tarkwin, with their sample bags and chipping Hammers, climbed up into the Himalayan's. They made observations and did a lot of chipping. One day Sheridan cried out 'WOW, Tarkwin look at this, Lava up here in these high mountains. This proves my idea that these mountains where created by upheaval and volcanic action". They raced down the mountain and Sheridan wrote a "paper" that he presented to a bunch of bearded antiquarian's, at the club. They considered it a logical answer to their quandary of how those dam mountains got there. They had Tarkwin rush it over to the publisher, who put it in the book we read." Every one lived happily ever after, end of story.

We would like now to speak about this under the Crust Magma Current that science is always talking about. They claim it can move Continents, they claim it can build Mountains. That's all a lot of Hogwash. How? In the name of common sense could anyone believe that a current could move North and South America to the West, India to the North, and Australia to the Northeast? They are also supposed to have moved Madagascar away from Africa towards the East. When not pushing Continents in every which direction these currents are also credited with combining with Volcanoes to, as they say, build Mountains. Nothing from underneath ever pushed up a mountain or even a hill. Current Magma flow under the crust of Earth is slow and directionally consistent. Being very sarcastic, (Ma said it was from Pa's side of the family). Let me say here that to make the currents under the crust do, what science says they do earth would have to be taken in the giant hands of God, and shaken violently. Like those little Christmas Globes with Santa inside and a lot of snow in the water. Shake them real good and the currents go in every direction.

Wow! As long as we're in this philosophical mind set, lets speak about Earthquakes. We've read about Tectonic Plates (which we do not question) and we've read of all these Faults that cause slippage of our crust and quake the Earth. We remember a few years ago hearing a radio report of an Earthquake in, we think it was Bakers Field, Calif. A learned Seismologist was invited to explain the phenomenon to the listeners. Since he knew there where no Tectonic Plates in the area and believing that quakes were caused by slips along fault lines he consulted his charts. "Lets see here, oh, we didn't know there was a fault here in Bakers Field", he said. Even though it was on radio we could see him reaching for his chart with his red pen and carefully drawing a new fault line through Bakers Field. He most likely had done the same thing before. Why didn't he say, "Wait a minute, there's no fault here, so what really caused this quake?". He didn't say that because his Seismologist Professor at the University taught him that Tectonic Plate movement, or fault line slippage caused quakes. Being thus educated, he did not use his intelligence to look for the truth. The truth did not compute with the Baker Field Incident, so he drew a new red line, Oh his Professor must be so proud of him.

Well all that aside, we must say that we have never experienced a quake. We have look at many pictures of the aftermath, and watch TV coverage of various incidents. One TV story was about a horrible elevated roadway collapse. The Engineer that was interviewed spoke about the construction of the roadway and the great strengthening of the structure to insure that plate slippage and the side to side swaying it would cause, would not in any way effect the integrity of the structure. Three levels crashed down on each other. It seems some sort of an upward movement had caused the disaster. Gravity had been considered sufficient to sustain the structure from any small vertical movement that might be encountered. No one seemed to ponder about how the side to side scraping of a fault could cause upward pressure. No One! (That may not be completely true. Some student out in Minnesota may have argued it, with his Professor, but the poor kid failed the course because he deviated from the written facts.)

We have also seen pictures of great cracks in the streets of an Alaskan Town, pushing one side of the street six or seven feet higher than the other side. "Hey Guys, Seismologists, Geologists, the above kind of damage could only be done by a tremendous whack on the belly of our crust". Those of you who know the story of Commander Joseph Hovar also know the source of this mysterious whack. A piece of the great rock about 1/4 the size Australia has been moving with the Magma Currents around and around our globe for eons. Speaking of these Currents. Science draws pictures of them moving from the North Pole to the Equator and also moving from the South Pole to the Equator - What a lot of foolishness. If you fill a kids toy called a "Top" with water and then spin that Top, the current of the water inside will move consistently in the direction in which the top is spinning. Of course you don't see it because you have a tin Top. Go out and get a clear plastic one and add some fake miniature snow to the water and the whole idea of Magma current direction will be made "perfectly clear" The water in the Top is thin and the Top spins fast, so the water and the snow move fast. The Magma in the Earth is thick and the Earth rotates slowly, so the Magma moves slowly and pushes the great rock even a little more slowly.

Let say the great rock hits the undersurface on its flat side and slides along for three minutes, before breaking loose and continuing on its way. Witnesses will say the quake lasted three minutes. Science assigns a number 4 on their scale. It moves along and on the other side of the globe it tumbles along and violently strikes the crust with one sharp corner twists away and collides again with another of its corners. Witnesses say there were two violent shocks about two minutes apart. Science gives it a 6 on their Richter scale. Now it could travel around four or five revolutions and sink deeper into the Magma and not strike anything. It is possible that the Moons Gravity could, in this case, pull it up towards the surface and it might make an enormously violent strike with its full surface, imbedding itself then breaking loose and tumbling along to the next incident. Witnesses will cry about their home being knocked off their foundation, glass breaking, ceilings falling, and the terror. Then the quaking again and yet another jolt. It will be assigned a Number Ten.

We often hear about the dreaded after-shocks. What are they, you ask? As we have said before this Rock or by now possibly two or three rocks have been traveling around for eons. Pieces have been breaking off and like a Comet have formed a Tail and follow the rock, striking where it strikes, and moving on.

What about volcanoes, you ask? Well Volcanoes are simply safety valves that were "placed" in various locations to relieve the pressure that builds up inside our sealed pressure cooker. If these valves had not been developed the pressure beneath us would have caused humongous explosions and they would have ripped giant holes in our surface and making vast areas uninhabitable, they are not natural. Nature could not deduce that without safety valves cataclysmal catastrophic problems would occur.

Along the bottom of the Pacific Ocean we have what is called the "Ring of Fire". This ring is a multitude of volcanic safety valves serving two purposes. The first is to release pressure under water and away from inhabited areas of surface land and humans. The water also filters the smoke and ash being expelled. Should all of those same volcanic safety valves have been placed on the surface our atmosphere would be seriously contaminated. Some safety valves are on land as pressure builds up in various places not just under the Pacific. Secondly volcanoes produce great heat. In order to have ocean currents or the movement of water, this heat works as part of the current machine. Hot water rises to the surface forcing cooler water down to be heated creating currents. Commander Hovar and his crew thought of everything.

Lets find a low bridge, on a calm day, over a small rectangular lake. Aha, there it is. Grab a small stone and that long bamboo pole and come out to the center of the bridge. Take the stone and drop it straight down into the lake. Well just look at those beautifully symmetric circles radiating out to the shore of the narrowest part of the lake and to the longest end of the lake, how beautiful. Now take your pole over the side of the bridge and swish it back and forth in the water. We seem now to have an egg shaped oval radiating out from the water disturbance. We suppose that if the Ocean floor were struck sharply from below beautifully symmetric circles would result, under the surface. And a Tidal Wave would rush towards the shoreline. Science would follow these symmetric circles to the shorelines of the Ocean and call them a Tsunami. We are not sure what would happen if a fault slipped under the Ocean or a Tectonic Plate moved. Would the circle be symmetric or egg shaped? (By the way, we were never able to really find that lake with a low bridge so we had to use our creative thinking for our experiment.)

Commander Hovar deemed it too dangerous to create another hole to be use in removing the rock and placing it in outer space or crashing it into Jupiter. He was asked if these Earthquakes would negate the continuance of the plan to create intelligent creatures to inhabit the planet. "No" he said, "the Rock is not now large enough to break through the surface again and the creatures we create in our image will just have to live with it." So you see we have lived with it over the eons, but not until now have we truly understood why the Earth sometimes shakes and quakes.

These have all been imaginative thinking stories or theory explanations. These stories are hypotheses. We would like you to SUPPOSE they might be true. We would like you to ponder them, to contemplate them. We would like you to wonder and say, You don't think, you don't SUPPOSE this could be true.

What we do know is that the Indian Sub Continent slammed into Asia after leaving the Antarctic. We know that we experience earthquakes. These stories, this theory can't be disproved. Arguments can be made that the kind of technology mentioned could never exist or that the hypotheses are not consistent with the presently scientifically accepted theories. But there is a way it might be proved. If we could set up Sonar Stations at various locations that were capable of seeing under the crust they might notice large objects moving slowly along in the magna. Those or that object would be the great rock and it might be tracked and in time give us the ability of advanced earthquake warnings.

"You idiot"...Oh, I'm sorry... that's a scientist who knows Earthquakes are caused by slippage between Tectonic Plates. He's quite sure of himself so I would rather not engage him in conversation, but if Commander Hovar was around we could have quite a discussion. Now that that scientist has left, we might explain our idiocy for you. Tectonic Plates are cuts in the Magma Crust that where necessary in order to be able to move landmasses across the Magma. After the Tragedy at India most of these planned moves where cancelled but the cuts remained and do from time to time experience slippage.

These original cuts in the Magma, when it was molten have tended to cure themselves when water was added to our orb. Though cured they are slightly unstable and when the Rock passes under them the pressure between the Rock and the crust does sometimes cause slippage. Pressures of our planetary revolutions and gravitational pull of the Moon also can cause disturbances.

We now leave it to you--what do you SUPPOSE?

By the way I can't be an Idiot, all of the above is just imaginative thinking.

Before the unsuccessful attempt was made to separate Europe and Asia a more successful rearrangement of landmasses was accomplished. South America was separated from Africa. At the same time Europe and North America were drawn apart. Science has helped us prove one of the methods used by the Creators in moving North America. Science calls it Continental “Drift”.

CONTINENTAL DRIFT

We believe that South America moved away from Africa, once it was separated from it, with many less problems than the Celestial Engineer Beings had moving North America westward. We spoke about the ability and technology of the Creators endeavoring to move Asia away from Europe in "The Tragedy". We spoke about their plans but did not go into any details about how they intended to move Asia. Let us now explain how the North American Project proceeded.

This will take a strong Open Mind. You must put aside other hypotheses about tectonic plates, faults and inconsistent under the crust Magna Flows doing a lot of huffing and puffing and pushing plates along with what science calls Continental "Drift". We will be in a Science-Fiction Mode controlled by our imaginative thinking. It could have happened this way and it is up to you to SUPPOSE it did, to ponder, to contemplate, and ask yourself, "Do you SUPPOSE this could be true?"

As you know, in the beginning, there was one landmass. Greenland was attached to Norway. Canada was attached to Greenland. In the beginning there was no water covering the Planet. Had there been water the solidified bottom of the land crust could not have slid across the also solidified top crust not covered with land but with water. Thusly the Creators calculated that the landmass could be slid and pushed across the molten magma. The continent separating cut between Europe and North America was the first cut of Project Earth. The application of a Black Hole Powered type of electro magnetic anti gravity beam cut between Greenland and Norway. Ireland, Great Britain and Denmark were tucked into the Bay leading to the Baltic Sea. The Cut continued past them and continued on by the Netherlands, Belgium and France to the point of the now known as the Pyrenees.

With the cut made three or four Star Crafts were assigned to push the separated landmass West. The subscribed distance of ten miles was accomplished. The ten-mile gap between Europe and North America of molten magma sagged and before they could slide it further the sag had to be corrected. A smaller craft located in the north, moved south over the gap pulling the sagging troth up with its electro magnetic beam. At the same time the craft cut into the center of the ten-mile wide slit so the next westward push would go more smoothly. Now with the troth level another concerted effort was made moving the landmass another ten miles west. The smaller craft using its electro magnetic power moved now from the south to the north leveling the troth and cutting into its center. Then again the push and the then north to south Electro-Magnetic pulling up of the magna leaving a series of five mile wide strips of magma.

Things were going good but then Greenland broke away from Canada. Because of its comparatively small size it was deemed to dangerous to apply pressure to it continuing its westward travel, as it could break up. It was left where it decided to stop. One of the spacecrafts had applied pressure in one spot for the entire procedure. The spot was that which we now call Hudson Bay. The usual practice when pressure caused an indentation in the crust was to pull the surface back to its normal position. Calculation in this instance showed that if Anti-Gravitational lifting pressure was applied the crust might crack causing a volcanic condition. The indentation was not disturbed and later it was filled with water.

Another craft had caused a large indentation in the area called the Adirondacks. This area was stable and so pressure was applied pulling it up and causing a dome area, which exists to this day. Further inland the crust was thicker and great pressure was used in Arkansas pushing the landmass west and the excessive pressure created diamonds in that area. The pushing pressure caused another gouged out area between Niagara Falls and the Wisconsin Dells. (Science tells us it is the same rock in both places). This area was later enlarged and proved to be a Great place to put Lakes to store fresh water.

While the movement of America was managed in the East other crafts in the west aided them by making the sliding easier. The westernmost crust was lifted slightly and other spacecrafts proceeded to sub-duct the Magna.

(Under normal condition subduction would be the proper Engineering procedure and there would be no problematic incidents.) Vice Commander Albert Lau caused a serious series of problematic incidents along the California Coast. Once the Great Rock he created was set loose the subduction became a severe problem. By now the Earth had been moved to its present position of Third Planet and the Magna was slowly cooling. If the subducted hardening magna hanging down along California's Coast was struck by the Great Rock, it could rip the coast apart. In order to at least minimize that catastrophe it was necessary to cut through the surface at various steep angles to cut off the Sub ducted material shortening and bringing it closer up to the surface and hopefully not deep enough to be struck. (Oh by the way, these laser cuts have been labeled faults by science and blamed for earthquakes rather than blaming Vice Commander Lau's error in judgment.) These laser types of cuts seriously chopped up the California Coast and later much mud was brought in to cover them and remove the look of construction from the land. Man has built homes and roads on top of these mud cover-ups and now when they get rain they get mudslides. Mud was also piled up in Central America to create a land bridge between the North and South American continents. This bit of construction by mud caused a great deal of problems for the French and the Americans while they each in their time worked on the construction of the Panama Canal.

During the initial ten-mile moves the sub-ducted material became entangled with the bottom portion of the crust causing a plow like pile of material that impeded progress. To alleviate this problem Star Ships were moved in. The crust was grasped by their power and pulled up causing great ranges of mountains some called the Rockies and at the same time leveling the crust bottom so the sub ducted material slid easily by.

South America experienced the same problem and the same solution was applied creating the Andes. Had you really thought that the magical ever flowing, powerful Magma Currents had pushed those Mountains up from the bottom? If you did this might be a good time to sharpen up your common sense and shake up your logical and reasoning powers. That pushing up business would have taken an awful lot of push-ups and lava would have been spraying out of cracks caused by its earth bending pressure.

After Greenland broke free and the westward push continued Northern Canada began to fragment due to the more sharply rounded shape of our orb nearer the North Pole. The Creators persevered and North America slowly moved into its designated position. At the time, enough space was set aside on the western boundaries for a new ocean to separate North America from the then proposed continent of Asia.

The knowledge of continents being moved ten miles at a time across the magma was given to us at SUPPOSE by our friends the Scientists. Fortunately for us science did core drillings off the coast of Iceland

Examination of these core drillings by Scientists established that small pieces of metal in the Magma in one five mile strip were facing North and in the next five mile strip faced South. This reversal of magnetic polarity every five miles indicated to the scientists that Magnetic North continually jumped from north to south during the "drift" of continental North America. Magnetic North today is slowly but yet continually moving around in an unstable condition. We would have to suppose that science feels that's just the way nature is. (That's always their answer to questions they can't answer) I imagine they feel that today's. unstable magnetic north is because it jumped around so much in it's youth.

SUPPOSE though has let all you wise, open minded, highly intelligent seekers of truth know that it was the Magnetic pull of a space craft on molten magma troths that caused all this phenomena. We have also wondered how science thought all these little pieces of metal could twist and turn north and south while imbedded in solidified magma deep under the water of "their" always-Blue Planet.

The principle of the polarity of the north magnetic pole changing itself back and forth did not and does not make sense. It does not qualify for scientific sense and it does not qualify for common sense. Let's look now at "common sense" and ponder on why science doesn't use it more often.

COMMON SENSE

I've always wanted to say but have never had an occasion to: "OK no more nice guy, from now on I'm telling it as it is." That's what this Section is about. I'm going to ask you, even if you are Scientifically educated to try to use your common sense. This will be a new experience for you as it is something that you never had to apply during your years in an institution of learning. Evolutionary education would be drawn into a condition of muck and mire if common sense where allowed to permeate it. There is also another condition that needs to be applied in this section and that is your intelligence. Intelligence, it seems to me, is another function that was never required when you were institutionalized and being educated. If it turns out that education is simply memorization of accepted ideas then naturally and truly intelligence and common sense would not even be needed.

Since I'm eighty plus years old that makes me very wise. So as a wise guy I beseech you to please use your intelligence and common sense and look deeply into evolution and try to comprehend the misinformation you memorized when you were in school. There is no "actual" connection between the fossil record and evolution. Even though the theory of Evolution was developed to explain the existence of the fossil record. Naturally science rejected Biblical Creation but since we did exist Darwin and some others dreamed up the theory of evolution to explain our existence. Then they welded that theory to the genetically engineered design change record of fossils. These design changes were made during the creation of the various creatures by Celestial Beings. Common sense will inform anyone that the complexities of living creatures could not have been accomplished by the spontaneous random joining together of cells. The fossil record is a record of design changes during creation and the theory of evolution denies design and denies creation.

The theory of evolution should never have been thrust upon mankind. True science should have said "Here is the Fossil Record that shows improvements in the development of living creatures over long periods. At this time we have neither a reasonable nor logical explanation of the cause and effect of this evolvement of fossils. As learned men of scientists we reject any proposal claiming that the complexity of creation could have happened spontaneously We shall continue to build the fossil record and study it with the hope of some day developing a cohesive theory that will explain it."

Instead of being honest, Charlie got together with his cohorts and dreamed up this unreasonable and illogical kindergarten thinking ridiculousness that was and still is taught to individuals who go to institutes of (get this) "higher learning" to be educated. (To claim that creature development scientifically just happened, like... Poof ... might be the answer one would get by taking a survey in kindergarteners.)

I'd like to be a nice guy but I must say that the theory of evolution is stupid. Even with that firmness of mind I realize that most people will think it's ME who is stupid. Now if you would, please gather together your intelligence and your common sense. Please push the fossil record aside out of sight and let's just and only look at evolution. Do you truly believe that asexual single cells could join together in such a manner as to create tissue and then continue on to create all of life's complexities? Complexities like those of fish, primates and other mammals, reptiles, insects, birds, flowers, trees, fungus, viruses and even you. Just imagine all these complexities with absolutely no control and with complete random haphazardness. Remember this joining together actually created (even though they don't like to use that word - created) digestive, blood and nerve-vision systems along with breathing apparatus.

Using common sense do you truly believe those cells could have made the thousands of decisions that would have been necessary during this creation of all things? Since the cells do not have any brains a little magic would have to be applied to evolutionary creation. Science and magic do not go together usually but in the case of the theory of evolution an exception must be made.

Science seems to think that hundreds or even thousands of little evolutionary change incidents or selections, occurring over thousands of years, explain creature development without design or creation. If Bob Hope was still around and told me that one I'd say "That's hilarious Bob, tell me another".

Since all religious Inquisition has been terminated I feel it's fairly safe for me to speak a bit blasphemously. So here goes... religions are business organizations whose product is morality. Guilt is their control, fear of the unknown is their hold and ritual their fascination. Remember now we are with open mind. Remember now that this is a fun book trying to humorously help us seek the truth. Please remember that we are asking "You don't SUPPOSE this could be true, do you?

The Judo-Christian religion has this book The Bible that is considered Holy as also other religions have their Holy Books. Holy seems to mean it has magical powers. We swear on the Bible in courts of law. We swear on it because we seem to believe we would be more likely to tell the truth holding our hand on it than we would with our hand on a pile of old newspapers. The Holy Bible should be revered for its antiquity and Godliness but it not magic.

The Old Testament contains God's record of creation and stories regarding early Hebrew history, Much of the Old and New Testaments are written by man. Both the old and new are interlaced with what religious business organizations believe our moral behavior should be. For centuries things have been added and things have been taken away. It's filled with man's ideas of morality. It's filled with educational teaching parables to get the organizations point across.

If you believe I've just said something disrespectful about the Bible you are mistaken. I am trying to say something truthful about the Bible. Its antiquity alone requires respect and awe. Its educational teaching parables have aided mankind as his societies advanced.

I believe there are lines that "man" wrote the words "and God said" followed by "their" written teachings. As long as it's taken with intelligence and common sense it's a wonderful book but it is not magic. It may be a best seller but it is not magic because we of the twenty-first century using our common sense do not believe in magic. Also there are those lines that say "and God said" that are the words of God. For God does live. Especially in the days of the Old Testament stories when God was walking among and guiding mankind in our development to being civilized creatures.

The Creators guided us by speaking to individuals like Moses and Ezekiel and having them pass their words on to those individuals' tribe. We don't hear a lot about God speaking to us nowadays. Unless of course we're some kind of kook. The Creators have pulled back and sort of left us to seek our own destiny. The Creators have untied the apron strings that they held us with for centuries during our pre history development. About six thousand years ago these Gods gave us writing and allowed us to record our own history for ourselves. Some of the Bible "corrections" made by translators and scribes have been to represent actual happenings as though they were just dreams or as they say visions, like in Ezekiel's case.

We grant you that from time to time things occur that we cannot explain with common sense. We call these incidents "miracles" and ponder their cause and effect. Someday we may understand miracles but now we can only ponder on them and know they are Godly. But remember that there are so many fabricated miracles one must use great gobs of common sense; contemplate with extreme determination and muster up all the intelligence available to us in order to ascertain a miracle's truthfulness. One must ponder if a coincidence or series of coincidences are just that or a miracle.

Sixty five million years ago was a time that needs a great deal of common sense. Two biggies happened then. First the dinosaurs became extinct then the slow process of evolution was shocked as birds "suddenly appeared". They tell us that a very big meteor crashed into Earth and caused such devastating effect on atmospheric conditions that the poisoned air killed off the poor dinosaurs all over the planet. It did not kill off every living thing, it was quite selective. Does your common sense believe a poison atmosphere could be selective? Dinosaurs were pretty big so this gaseousness must have been pretty powerful and also very magically selective.

Those of you that can't accept meteor explosive gases as the killers of the dinosaurs have other choices. It could have been an infectious plague, a sudden drop in temperature caused by a supernova, or locally a sudden severe sandstorm, or newly evolved mammals may have eaten the dinosaurs' eggs. Those are all very scientific hypotheses. You don't SUPPOSE that a lowering of testosterone without the availability of Viagra might possibly have been the real cause of the Dinosaurs Demise, do you?

Dinosaurs had some purpose for being created. It could be that they themselves were gaseous. They might have been created to help build the layers of atmosphere that surround and protect us. They could have been helpful in fertilizing the soil. Whatever their purpose there came a time when they were no longer needed. It had been determined that they should be replaced by new and smaller "flying" reptiles called birds. There would have been an order proclaiming the cessation of the dinosaurs' experiment. This termination of that experiment and other selective "extinctions" would have been carried out.

When reading an encyclopedia on prehistoric life (which I hope you do) it is a surprise to run across the statement of how this or that experiment of nature became extinct. Random does not experiment. Haphazard does not experiment. Spontaneous does not experiment. Natural living (celestial) beings do experiment and do make decisions on termination of experiments. There were no extinctions in prehistoric history, there were only the termination of experiments. Not until man came along did creatures "become" extinct.

Oh my God, how could birds not have evolved over long periods of time. How could they have suddenly appeared? The Creators here on earth were trying to create a flying creature using the dinosaur's protoplasm as a base. They made a lighter gliding creature with wings of skin and hollow bones but had trouble creating "feathers". The problem was put to the celestial beings at the universities in the Kingdom in Heaven and feathers were developed. Feathers were made in all colors and creatures built with phenomenal aerodynamics.

The shock of birds suddenly appearing was troublesome enough to science but when they did suddenly appear they shockingly also demonstrated a great leap in evolutions slow and steady development. It seems these birds not only had lungs and a four chamber heart but also a new development, their lungs had many air sacs.

The wings exercising to stay in flight and climb would cause the lungs alone to experience breathing problems. The altitudes reached would cause additional thin air problems for breathing. Science noted this great leap of evolution in the development of air sacs to aid breathing. Science accepted this development as common sense and a "natural accidental" step of evolution. The leap in this improved system was a little troublesome and difficult to understand for them.

We of SUPPOSE know that before an actual bird existed the DNA Code plan of air sacs were created to compensate for thinner air at higher altitudes and for constant exercise of wings for propulsion.

Earths environment was stable and suitable for birds sixty five million years ago so thousands of birds in suspended animation and eggs were brought to Earth and in laboratories hatched, released and "birds suddenly appeared".

Even though science believes that all creatures, including birds came into being by an accidental bumping together of cells in an uncontrolled haphazard and random manner they have been studying bird flight to aid our aerodynamic specialists in building better designed wings. Hey wait a minute scientists don't you remember birds wings were not designed, they just happened ...Like "Poof" you're a wing. Even though these aerodynamic specialists didn't know what they were doing they were studying the unbelievable genius of the Creators who knew more about aerodynamics sixty Five Million years ago than we may ever know.

I remember when my mother used sophisticated Twentieth Century common sense. It was the day she had watched our first astronaut being shot up into space in a rocket. When I came home from work she said "That's where I always thought Heaven was. I guess Heaven must be further out." Just a little common sense and she readjusted her seventy-year belief in the location of Heaven.

I did mention before this little common sense statement but it is well worth repeating. "The only thing you can be positive of is that you can't be positive of anything." Here are some more words of wisdom. Non scientific people who write on scientific subjects should keep in mind that "A little knowledge is a dangerous thing."

Here's another little common sense item for you. The theory of Evolution is a Pseudoscientific Theory. To bad no one told those Pseudo scientists that one could look at a series of circumstances and come to the wrong conclusion. Boy I'm glad I got it right.

It was never my intent to insult all educated people. Not being educated I can't be sure of all college courses so lets confine our insults only to those intelligent students who sat there in class and became educated in evolution and never said..."Wait a minuet Professor that doesn't make sense... It's not Logical...It's not Reasonable...There's no cause and effect...It just does not compute in my mind...My brain knows that the random coming together of cells could not possibly have created man."

(PAUSE)

Now that I've given the above some thought I must agree with the thinking process of the intelligent student, who kept his mouth shut. His parents spent a lot of money getting him into that school and it seemed imprudent for him to tell his Professor the truth and thereby chance not Graduating.

I believe it may have been Shakespeare who made a comment on students keeping their mouths shut in evolution class thereby insuring their graduation. What he said was "Oh, there's the rub that makes evolution so long of life, for what student would bear expulsion from class rather than keep his mouth shut and guarantee his parents investment."

To bad we hadn't been speaking of EGO, or I could have brought up man. Boy, what a big head we have. I've read over and over how man had domesticated animals. Man's supreme intelligence had designed the "wheel"; Man developed a system to create bronze; Man made looms; Man made Art; Man invented writing. All this great stuff man did. Unfortunately it was all accomplished by "man" during the period that man did not record his history. Yes man has been given credit for a great deal of development during his pre-history period. Wherever science found improvements in pre history "their" Ego forced them to credit this advancement to pre-historic men.

Domesticated animals had been genetically engineered to be domesticated by the Creators and then given to man. The wheel of course was Gods gift to man. Celestial Beings showed man the plans and then assisted him in building this labor saving device. Native Americans did not have the wheel because God did not choose to give it to them, not because they couldn't think of it as the wise Europeans had. The kiln was also given to men and he was shown how to smelt ore and produce bronze. Looms were presented to man along with the knowledge of spinning yarn and weaving cloth. The first art classes were in the cave mans time. The Creators knew that as there was no written history man would later be given credit for his ingenious intelligence and the FACT that God had walked with man would be kept a secret. As the time for the Creators to sort of leave us to our own destiny neared they began to pull back into the shadows. In private they taught privileged scribes a written language that those scribes could pass on to others. The scribes were taught that the appearance of God was never to be written or spoken of. The next generation of scribes never knew what the appearance of God was, the secret was kept.

We all now know that evolutionary scientist (as if there could be such a thing) need to admit common sense into their science. There is another science that also could use common sense to improve the understanding of their endeavors in science. This other science is Behavioral Science of both man and animal. The sexual nature of man and animals have been misdiagnosed since the Church took the soul away from science and left science with only the body and the brain and its Genes to contemplate and ponder, in their endeavors. Behavioral Science is getting nowhere in their study of the domestication of wolves. They have brought up wolves in a home environment with other puppies that were dogs. As they grew the dog puppies developed into dogs and the wolf puppies developed into wolves. Oh yes, the scientists did pondered, they did contemplate but they failed to see the whole of their science. Behavioral Scientists only contemplated and pondered on half of their experiment. They did not even consider the soul of the wolf. The Creators genetically engineered the spiritual essence that would be installed into a wolf with that which would make the body and brain of a wolf into a wolf. Behavioral Scientists "know" or they think they know that since man domesticated the dog, they could domesticate the wolf. The difference, which they cannot understand is that the dog, was engineered to be a dog, the wolf to be a wolf. What a leap in understanding we would gain if science were to allow their practitioners to look into the whole of their experiments instead of forcing them to only observe half of what they experiment with. Common Sense dictates to the intelligence that there is more to man and animal than just a heap of protoplasm. When it comes to the spiritual essence within living beings, religion has conspired to conceal this fact from science. That there has to be more to man than just the flesh science has ignored. The time has come for science to become, shall I say, Religious. The time has come for science to look with what their brain has already noted, that there "is" more to man and animal than just the flesh. The time has come for Behavioral Scientists to look at both halves of their experiments.

As I was recently perusing the Book of Genesis with my high degree of Common Sense I began to contemplate all the measures God took to convince the Pharaoh that he should let Moses people go. God turned the waters to blood and caused it to stink, inundated the land with frogs and then with hoards of locus. What unbelievable and miraculous events they must have seemed at that time to those people. I pondered for a while and thought of a rather un-miraculous similar event that has often happened in the New York area. This modern event we call "The Red Tide" which causes rivers to appear red and the water to stink with dead fish. I've also heard of thousands of frogs on the move, causing havoc on highways right in this country. Everyone today knows of locus swarming here, there and everywhere from time to time. My common sense says how strange it is that these events of nature that happened in the Bible are miracles and when they happen in today's world they are nuisances.

I now sense a dark cloud moving in over my being. I feel a sadness moving into my soul. I hear a voice saying, “It’s time to go”. This all must come to an end. This search for truth must end and be left to supposition, contemplation and pondering. This all must be concluded.

IN CONCLUSION

How awful to write “in conclusion”. That means it’s all over, it’s ended, and the fun has stopped...How Awful. What will I do with my time, what will we do with our time? ‘We’ being, my cosmic being and my biological being. Words and Phrases have jumped into my fingers as I typed. Thoughts and ideas have leaped into my consciousness from somewhere outside of myself. It has all been truly fun. We’re not finished yet, we will add some more thoughts here and there and drag on conclusion as long as we can.

I remember there was an “In Conclusion” in Darwin’s “Origin of the Species”. I wonder if he felt sad on writing it. Probably not, as he was in a hurry to get credit for his thoughts and wasn’t really enjoying his exercise in imaginative thinking. Another thing I remember about the “Origin” is that some kind of a tree or bush was growing up here in North America and was also found in South America. The conclusion to how that happened was that a bird ate the seed of the plant and then moved his bowels on a log that got frozen in a glacier floating south and melting. The log thus released from the ice floated ashore and the seed washed up on the soil and planted itself. When I read it was hard to realize that anyone could write such hogwash.

SUPPOSE on the other hand would conclude that Harvey, working on a beautiful September morning in the north had wondered how his plant would react to jumping from Autumn to Spring, without a Winter. Harvey would have called Stanley who was a fellow Botanist, working in South America. “Hey Stan, I’m going to fly down this afternoon with a specimen I’d like you to observe for me, and I would like to pick up some plants of yours that have just completed their winter dormancy to see how they react to a second Winter.” How simple, how logical, how understandable. It’s true that SUPPOSE ‘s conclusions might be more understandable but unfortunately they would not be as scientific as Darwin’s would.

Reincarnation is a fascinating subject. When you believe in it you can’t help wondering whom you might have been before. That wondering would be very hard to trace down due to the millions of people dying and being born each day. Hinduism had a class system and so they believed in your soul being reincarnated into different classes. We of SUPPOSE believe this soul transfer is into the same model from which you were originally cloned. Mating with different models over thousands of years would of course draw those models into the stew.

Lets take a look at someone famous like Franklin Delano Roosevelt. He died in April 12, 1945. He had an outstanding charisma. He was a Democratic Governor of New York. He was a womanizer, a dog lover, a stamp collector and a political socialist who often said “I love being President.

Then there was another President born on August 19, 1946. He had an outstanding charisma. He was a Democratic Governor of Arkansas. He was a womanizer, a dog lover, a stamp collector, and a political socialist and often said “I love being President” His name was William Jefferson Clinton. This does nothing to prove reincarnation but it is interesting to note.

Thomas Woodrow Wilson was a peacemaker. He wanted to have WW1 be the war that ended all wars. He wanted the League of Nations to guarantee world peace. He wanted peace in our time. He was not successful. He died on February 3, 1934.

On October 1, 1934 another President was born. He was also a man of peace. He tried in vain to bring together the governments of Palestine and Israel. Even after his Presidency he traveled to those lands seeking peace, even though told not to go by the State Department. His name was James Earl Carter Jr.

Another series of coincidences that might draw you to a correct conclusion or lead you to a misguided belief. It is though also interesting to note.

Before I conclude I’d like to tell you of a conclusion I came to unilaterally and arbitrarily. There came a time just before the celestial beings left us on our own (more or less) that there was a movement among them to leave a record of the work they had done while constructing the Earth, and creating all things. At that time Moses and his people had developed a stable and reliable society.

Moses and his people were then “chosen” by these Celestial Beings whom Moses called Gods to receive and care for the record of creation. These Celestial Gods gave man the five Books of Moses. Genesis was written in an evasive manner but also showed what the conditions on Earth were like when they arrived from outer space. It described the necessary adjustments they had to make. They spoke in days instead of eons to accomplish their deeds and to come within the understanding of these technically unaware people. They told us that the Earth did not revolve. They gave Moses the story of Adam and Eve as the first of his cloned people to copulate. They told us all about the longevity of the soul. Moses knew of the appearance of Celestial Being Gods but met with them privately and away from his followers. The closest description we have of God from the Bible is that we are made in his "image" and the statements that they had the appearance of men.

All races and all nationalities had an Adam and Eve. All races and all nationalities had souls installed in clones. But only Moses' people were chosen to receive a written record of these events.

Before we leave you with our thoughts, which we fear may be lost in oblivion, we would like to tell you why we put those thoughts into writing. We felt it incumbent upon us to pass along to our fellow man this new concept of an actual living breathing being called GOD. A God "being" that you could believe in with simple understanding, simple logic and simple reason. Belief in God should not be complex or confusing.

We feel that man is passed his time of superstition and of his fear of the supernatural. Everything in our being denies supernaturalism, everything in our wisdom denies supernaturalism, everything in our logic and reason denies supernaturalism and yet we cleave to it when it comes to religion. Change is very difficult, almost impossible when it involves belief in God. But it is time for that change; it is time to consider a different concept of the same God.

The fact that this living breathing God is an extraterrestrial makes your acceptance thereof even more difficult. We have given you some examples from the Bible that the appearance of angels, lords, and the sons of God were different and even eerily strange. The Bible notes the difference between man and alien. The logical and understandable system of creation by design employed by the Celestial Beings proves the existence of God the Creator by illustrating the cause and effect of creation. When this system of

Designed Creation is scientifically applied to evolution the randomness and haphazardness of its theory fade to nothingness.

We would like to SUPPOSE that in the future the Designed Creation and Genetic Engineering procedures of Celestial Beings are taught in High School and college science classes in place of evolution. SUPPOSE hopes that this same subject of creation by Celestial Beings, by design and using Genetic Engineering is also taught in Sunday and Theology Schools in place of supernatural creation.

Science and religion must be drawn together because only together can we live with truth and understand. Only together can mankind accept the doctrine of Creation by Design. Creation by design and the Genetic Engineering of God the creator have to be recognized by science. The "science" of Designed Creation by Genetic Engineering has to be recognized by Religion.

Thank you for reading my words,
giving thought to my thoughts,

SUPPOSING, PONDERING, AND CONTEMPLATING.

With a tear in our eye and a lump in our throat we now again say, "Thank you" and "Good by".

the end...?

www.ingramcontent.com/pod-product-compliance
Lightning Source LLC
Chambersburg PA
CBHW021943170526
45157CB00003B/904

* 9 7 8 1 3 1 2 0 3 2 8 8 0 *